PUNCHINGCLOUDS

LASSE GERRITS

PUNCHING CLOUDS
AN INTRODUCTION TO THE COMPLEXITY OF PUBLIC DECISION-MAKING

Emergent Publications
3810 N. 188th ave, Litchfield Park, AZ 85340, USA
www.emergentpublications.com

ISBN: 978-1-9381580-0-1

Layout, design and artwork by Zeno van den Broek
www.zenovandenbroek.com

Library of Congress Control Number: 2012944750

Printed in the United States of America

Chapter 1:
COMPLEXITY AND PUBLIC DECISION-MAKING

Chapter 2:
BEING DEPENDENT: COMPLEX ADAPTIVE SYSTEMS

Chapter 3:
BEING DYNAMIC: BETWEEN INERTIA AND RANDOMNESS

Chapter 4:
BEING HUMAN: COPING WITH COMPLEXITY

Chapter 5:
COEVOLUTIONARY PUBLIC DECISION-MAKING

Chapter 6:
BEING INQUISITIVE: RESEARCHING COMPLEXITY

CHAPTER 1.

COMPLEXITY AND PUBLIC DECISION-MAKING.

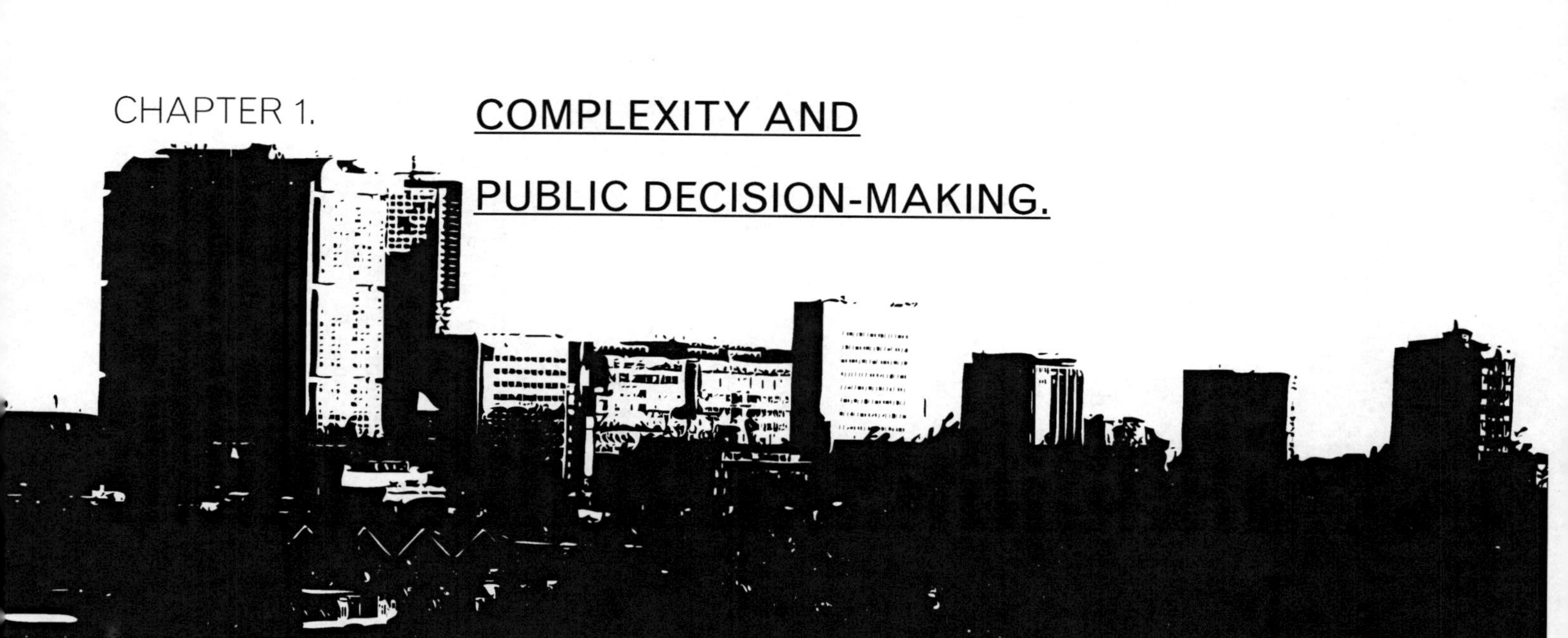

1.1

THE COMPLEX PART OF PUBLIC DECISION-MAKING

Much of the work in the public sector is fairly simple, unchangeable and predictable. A minor part of the work isn't. Although seemingly small, this complex portion requires much of peoples' time and energy, and presents often unpredictable results. A fatalistic response to this complexity could be to give up and to go home. A different response would be to make an effort at understanding this complexity. This book presents such an attempt. 'Complexity' is an amorphous concept that often equals 'incomprehensible' in daily parlance, but this does not need to be the case. Complexity can be named and analyzed. Once understood, it becomes possible for one to cope with it more effectively. From this perspective, the complexity of public decision-making mirrors the intricate beauty of the lives and interactions of social beings and analyzing the complexity will help us to understand the inner workings of that beauty. But such an aesthetic argument may not be convincing to every reader, so we first have to an answer to a short and simple question: why bother?

The realm of public decision-making is littered with examples of decisions, policies, projects and programs that should have worked, 'if only...'. For example, a study on the progress and pitfalls of large-scale infrastructure projects instigated and paid for by European governments found that a majority of such projects were not completed on time and experienced cost-overruns (Wetenschappelijke Raad voor het Regeringsbeleid, 1994). The report concluded that these projects could have been finished on time and on budget if some idiosyncratic events had been avoided or were better handled. Public Administration as a scientific domain has a strong heritage of developing recommendations for any type of public decision-making in order to avoid the pitfalls of previous policies and projects. However, the nature of the world causes public decision-making to be subject to unforeseen, unintended and sometimes unwanted effects despite the most well-intended of recommendations. It is good to consider and reconsider the way decisions are taken, but one should not believe that there is a perfect way to achieve faultless and uncompromised decision-making. No matter how one looks at the world. and no matter how we feel about it, the world is an erratic place and is full of both welcome and unwelcome surprises.

Let's return to the example of a large-scale infrastructure project. The perceived failure of such projects is important because they are often prestigious and it can cost politicians dearly once they start to run out of control. There are obvious incentives to keeping them manageable. Flyvbjerg, Bruzelius and Rothengatter (2006) have looked into the reasons why such projects are prone to failure. They find that "[...] most appraisals of megaprojects assume, or pretend to assume, that infrastructure policies and projects exist in a predictable Newtonian world of cause and effect where things go according to plan. In reality, the world of megaproject preparation and implementation is a highly risky one where things happen only with a certain probability and rarely turn out as originally intended." Because the stakes are high in such projects, there is considerable pressure to paint rosy pictures and to be overoptimistic about the results. The pressure is such that deliberate miscalculations are often made to mask the uncertainties and risks inherent in such projects. In other words: analysts, officials and consultants have a tendency to tell lies in to keep the projects looking attainable (Flyvbjerg, Skamris Holm, & Buhl, 2005; Flyvbjerg, 2009; Næss, Flyvbjerg, & Buhl, 2006).

While disappointment is often the result in such project and policies, some causes of disappointment are in fact repairable. Lying consultants and cheating officials can be fired and attempts can be made to take away the pressure to lie. But the problems are unlikely to go away completely. Looking at specific projects more closely, one notices that each of them is embedded in a specific context with specific local conditions impacting it that are unique to that particular project and rarely observed in other such projects. Another impacting factor is events that occur elsewhere in time and place. An example can be found in the Betuweroute project in the Netherlands which concerned the construction, with public funds, of a dedicated freight railway between the port of Rotterdam and the German border. As with many other such projects, the construction turned out to be much more expensive than planned and was delayed by many years. Looking into the causes of the time and budget overrun, one will find that its development was not only impacted by unforeseen local conditions, but was also impacted by dynamics occurring elsewhere. In particular, a series of accidents that occurred in tunnels in Switzerland and France around that time caused local authorities and the Dutch Parliament to place pressure on the planners to redesign the safety systems of all tunnels in the track. This, and other incidents, resulted in severe delays and budget overruns of approximately 800 million Euros (Algemene Rekenkamer, 2003). The Court of Audit blamed the Ministry of Public Works for not being clear about the uncertainties in the project and the Minister had to defend herself against widespread criticism. This example demonstrates how seemingly straightforward projects can be severely hampered by events far removed from the realm of control of the people managing them. The people in charge of the project or their advisors could be fired but that wouldn't solve the dynamics inherent to this capricious world. If the world is a dynamic one that is imperfect, people who would like to get things done should not be attempting to 'fix' that imperfection, but to work around it.

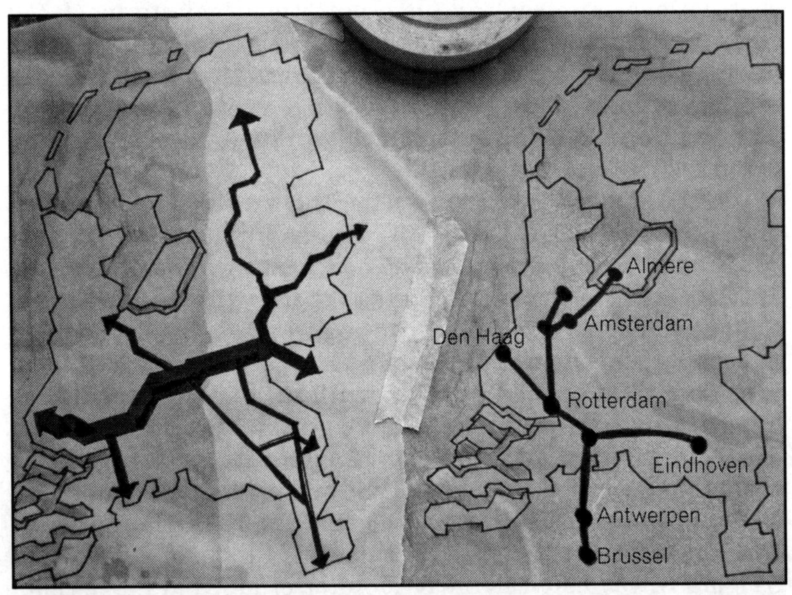

Large infrastructure projects such as the Betuweroute (left) and the HSL Zuid (right) are very difficult to manage because of their internal and external complexity.

Public decision-making concerns decision-making in systems that are not passive but that respond dynamically to many incentives, sometimes forcing decision-makers to take a reactive position against their will. Society and its many components will elude control far more than that it will respond to steering incentives. In some cases, the results of earlier steering attempts could actually become impediments in later such efforts (Gerrits, 2010; 2011). Public decision-making is often like punching clouds: considerable energy is put into the punching but the cloud goes its own way regardless of the punches.

The goal of this book is to name, analyze and comprehend the complexity of public decision-making. This requires an understanding of the very concept of complexity, an understanding of societal complexity and an understanding of how public-decision makers operate within this complexity. In particular, this book sets out to answer the following question: what explains the complexity of public decision-making? To answer this question, this book draws from insights and theories originating in the realm of the complexity sciences. Many of those insights and theories originate in domain of natural sciences, such physics, chemistry and biology. Some of them are around for almost a century whereas others are from a much more recent date. Scientists from diverse backgrounds such as Stuart Kauffman, Ilya Prigogine, Murray Gell-Mann, John Holland and Brian Arthur, among others, built on those works and started to (re-)combine them to form a body of (sometimes loosely) related theories that is commonly grouped under the header of complexity sciences. The complexity sciences gained popularity in the 1980's as increases in computational power allowed for much finer-grained research that impacted then-dominant assumptions in science such as time-symmetry (Waldrop, 1992). New theories, concepts and vocabulary that the computational science allowed then seeped-through the social sciences in general (e.g. Byrne, 1998) and Public Administration in particular (e.g. Mathews, Michael, White & Long, 1999; Haynes, 2003), where they incited novel ways of thinking. The new computational abilities also prompted the rediscovery of earlier concepts and theories that had pointed to similar phenomena before the scientific methodologies existed with which to test them. Authors pioneering the use of complexity sciences in Public

Administration include among others Kickert (1991), Kiel (1989), Morçöl and Dennard (1997) and Haynes (2001), with more advanced contributions coming recently from e.g. Room (2011a; 2011b) and Morçöl (2012). These are important works for those familiar with the complexity sciences but slightly out of reach conceptually for the beginner. This book is born out of the desire to serve readers who are not intimately familiar with the complexity sciences but who would like to understand what it has to offer to the understanding of public decision-making. It offers a bridge between the study of public decision-making for public administration and public policy on the one hand, and the complexity sciences on the other hand. As such, it tries to shed some light on the complex nature of reality and serves as an introduction to how and why this matters for public decision-making. Note that the subtitle of the book says '*an* introduction'. This description was used expressedly to qualify our attempt and to modestly acknowledge the limits of what be explained about something so overwhelmingly complex as humans operating in the real world.

1.2.1 WHAT IS COMPLEXITY?

The word 'complexity' is often both cherished and feared. But what do people mean when they use the word 'complexity'? There are, as Byrne (1998) notes, several accounts of complexity which have both distinct features and features that are shared across definitions. In common conversation, the words 'chaos', 'difficult' and 'complexity' are often used interchangeably (Page, 2008) and 'order' is seen as the opposite of these words. However, this does not necessarily help clarify these notions. Let's first have a look at how the words are used in everyday parlance.

The adjective 'complex' is usually used when one encounters something that is difficult to comprehend, such as a five-page application for a tax rebate. However difficult it is to deal with such things, they represent order as their mutual relations are fixed and the outcomes are predictable. Such difficult but ordered and structured things will be referred to not as 'complex', but as 'complicated' in this book. 'Chaos', is similarly used in common parlance to describe complicated things that people do not like, such as a trying to get through a crowd at a station in order get on the right train home when it is rescheduled because of an accident somewhere in the network. Negative and confusing such disruptions may be, 'chaos' itself does not necessarily have a negative connotation. As a scientific term, chaos is not the absence of order, but rather the impossibility of prediction because the outcome is sensitive to minor variations in the initial conditions. This mathematical take on chaos is founded on non-linear equations that generate random behaviour and that are insolvable when taken apart in sections (Gleick, 1987; Kiel 1997). Although some terms and general ideas from chaos theory are borrowed, the mathematical concept of chaos itself will not be used in the remainder of this book.

Complexity, then, is neither difficult complicatedness nor annoying chaos. Both order and chaos emerge from the same type of systems described above, i.e. systems that are of a stable nature. For example: standardized bureaucratic procedures are very systematic and the right outcome is obtained by just ticking the right boxes or following the right steps. Although the task of completing the procedure may be onerous, its stability ensures a predictable outcome. In more popular works, complexity is sometimes defined as the boundary phase between order and chaos where stability and randomness are entangled in a tense state (Waldrop, 1992). However elegant and tempting this definition is, it is rather difficult to relate it to the real world of public decision-making. It first requires the state of systems to be observably orderly, chaotic or complex, and then requires that one can define and observe a boundary phase – whatever that may be. No solid empirical research has ever demonstrated where and how this boundary phase occurs. So we need to flesh out the concept of complexity more thoroughly.

A good way to dissect the notion of complexity is by first describing what is observed without (yet) analyzing it. For this we need to modify an example from Rescher's work on complexity (Rescher, 1998). Let's just assume that the world around us and in which we make decisions could be described using the following series: "abababababab..." This hypothetical series shows that the world consists of two elements ('a' and 'b') that are unchangeably related to each other ('a-b'). The two limited elements make any description simple and we can describe the rest of the world without actually having real knowledge of it because the series is repetitive ad infinitum. In other words, it has clear predictive capacity.

Of course, the real world does not consist of just two elements. Not only are there many more elements, each comes in multiple shades or subtypes. Even more so, elements are not necessarily from the same class so we need to supplement the alphabets with, for example, numbers (that also come in different shades or subtypes). In other words, we need to expand our series to describe the world more accurately. First, we add more elements: "abcabcabcabc..." Then we add the different shades of each element: "aabbccaabbccaabbcc..."; followed by elements of a different class: "aab1bccaab2bccaab3bcc...". Finally, we recognize that the elements do not necessarily come in a fixed order and change the series accordingly: "caa11bc763ccz9xbvvvc33c2990a0abbcc..."

The point of this little exercise is to show that the real world consists of multiple elements, from different classes that are related, but sometimes loosely, and whose mutual relationships are changeable over time. We have no means of knowing when we have extended the series enough to describe the whole world. To be more precise: to describe the real world accurately, we would need a descriptive series that is as large as the world itself. Since we know that mutual relationships between elements can change over time, this series would need to change accordingly. Thus, the complexity of this world arises from the fact that the world is an enormously diverse place where local interactions between elements always render new and different outcomes. This simple statement has considerable implications for how the world operates, how we can know its operations and how it can be governed.

1.2.2 WHY IS THERE SO MUCH COMPLEXITY AND WHY CAN'T WE STOP IT?

Ostensibly, complexity hinders public decision-makers and their work. If the world consisted of only a few elements with stable relationships between them, it would be a predictable world and the consequences of each decision could be mapped in advance. It would operate in the same way as the tax return form discussed in the previous sections. However, any decision-maker in the public sector knows that this is not how the world behaves. There are ample complaints that complexity hinders sound public policies and there are continuous calls for simplification in procedures, laws, content etc. Why is it then that is there is so much complexity in public decision-making and why can't it be stopped?

The basic answer to that question is that complexity is self-propelling: it generates more complexity in an unstoppable way. The continuous interaction between the elements discussed in the previous section generates novel outcomes. And, as the examples from Europe's major projects show, chance or coincidence plays an important role. Unforeseen things happen, resulting in unforeseen outcomes. In a passive world, it just stops there. However, the real world is not passive. The social realm is populated with humans who feel compelled to respond to changes. Some changes are favourable, some changes are unfavourable and people may want to promote certain things and stop others. This response provokes new reactions that require yet more responses (Gerrits, 2010). Some authors writing about the workings of administrative and organisational systems have adapted Ashby's law of requisite variety (e.g. Beer, 1985, in Flood, 1999; Kickert, 1991) which states that for a system to cope with its environment, it needs to mirror that environment. Because that environment by definition is more complex than the system, it is inevitable that the complexity within the systems increases accordingly. A reinforcing loop between the system and its environment thus leads to more complexity.

An illustration of this phenomenon of self-propelling complexity is the proliferation and ensuing "hypercomplexity" (Rescher, 1998) of laws and regulations. Each new unwanted instance is often met with regulations aimed at avoiding a reoccurrence. Automobile safety regulations are an example of this. Governments sought to improve road safety by demanding that car manufacturers build safer cars. As cars became safer, pedestrian deaths in fact increased because drivers started to drive more recklessly (Peltzman, 1975). In response, governments designed another set of regulations to make cars safer for pedestrians, redesigned roads to induce safer driving and penalized road violations more severely. While regulations covered basic issues and safety improved, the outliers or accidents caused by rarer conditions became more prominent in the accident statistics and new regulations were devised to cover those conditions as well. In other words: actions resulted in new circumstances that demanded new actions and the degree of complexity increased in step with these changes. Nowadays, designing a car or a road is as much about following the many regulations as it is about using the designers' imagination.

The result is a proliferation of regulations and institutions (Norgaard, 1984; 1994) that are increasingly difficult to govern and to keep under control. Recalling the exercise in Section 1.2.1, it appears that the more we know about the world (i.e. the more refined our series of elements describing the world becomes), the more we will notice finer-grained details (i.e. we will notice ever more 'a's, '*b*'s and '1's and so on), and the more coping mechanisms we will develop for those details (such as regulations). Fuelling this proliferation is a basic human condition of coping with uncertainties and perceived risks, with which we deal later on in this book. The important premise for this book is that complexity is not a choice, it just *is* - it is self-generating.

1.2.3 WHAT BASIC APPROACHES TO UNDERSTANDING COMPLEXITY ARE THERE?

The idea that complexity is inherent to this world implies that it is a property of both the physical and the social domain. This is true, but it is (unsurprisingly) more complicated than just that. Thinking in terms of complexity and systems has a long history in both the social and natural sciences. It is necessary to trace some of the roots of complexity studies in order to understand the position of this book. The most recent wave of work in this field is often referred to as complexity theory (Byrne, 1998). Theory, as it is used in this book, refers to a set of mechanisms that help explain a certain outcome (e.g. De Haan, 2010) and the use of the term 'theory' to describe that most recent surge of complexity-informed literature is, in fact, a subject to much debate as it concerns an amorphous collection of theories that draw upon many fields for their core concepts, including evolutionary biology, thermodynamics, self-organization, game theory, cybernetics and catastrophe theory (Goldstein, 1999). The complexity sciences thus use a large variety of related, but not always coupled concepts that help unravel complexity and explain bits of it but there is not one unified theory or theoretical framework that explains everything. We will use those ideas and concepts in this book to get a better hold on the complexity of public decision-making without trying to unify it all in one overarching framework. Such is the nature of social complexity that if it is studied on a particular level (for example on the actor-bound level), details of other levels (such as the social system as a whole) blur a little - and vice versa.

Research into complexity can be broadly categorized as having two main approaches: general or simplistic complexity and situated or complex complexity (Byrne, 2005; Buijs, Eshuis & Byrne, 2009). General complexity is built on the premise that, ultimately, the world operates according to a universal regime or set of rules that applies to all phenomena, regardless of their nature. "Where we can map such generic behaviours onto molecular, organismic, neural, psychological, economic, or cultural data, we may have found the functional universality class needed to explain phenomena in these areas of chemistry, biology and the social sciences." (Kauffman: 1993: 404). Essentially, this viewpoint focuses on the emergence of structures and processes depending entirely on the (fixed) variables within the system. Many of the archetypes of complexity theory that are often referred to, such as the computational models by Reynolds or Langton (Smith & Stevens, 1996), are examples of general complexity that have also gained popularity in the social sciences (Byrne, 2005)

General complexity is functional in demonstrating the principles of complexity. At the same time, it has a limitation in that it can't be ascertained that the observed rules are indeed universal. To understand this, we need to recall the series as discussed in Section 1.2.1. We found out that if we want to describe the world accurately the series would have to be as large as the world itself. The human brain, however, doesn't have the capacity to process so much information. As observers, we can only see and comprehend a certain portion of the series, say: "caa11[bc763ccz9xbvvvc3]3c2990a0abbcc...", where the section in between the brackets is the portion that is actually observed and taken into account. The issue here is that the remainder of the series is of paramount importance to explain the full complexity of the thing that is being investigated or experienced. It means that certain apparently (near-) identical observations could stem from mutually different configurations of local conditions. In other words: similarities (if found) may not indicate a generic mechanism, but rather sets of different conditions bringing forth apparently similar results. For example, both the Betuweroute project mentioned earlier and the HSL-Zuid high-speed railway project were carried out in approximately the same period and had similar delays and budget overruns. However, these project outcomes had their roots in different configurations. This hints at the fact that the social world requires a specific approach to understanding its complexity.

In social reality, the number and nature of the elements defining an emerging structure or process is not fixed, but changeable. Social reality is open, its constituent elements are connected ad infinitum with other elements through other elements, and the observed phenomena do not define their borders unambiguously (Byrne, 2005). What constitutes, limits and explains a complex development depends on the observer's point of view. It is the human observer who draws boundaries in complexity in order to make sense of it (Cilliers, 2001). Therefore, social complexity arises not only from the constituent elements of a system but also from the fact that this constitution is dynamic in itself, i.e. that it is constantly changing. An important idea behind the complexity sciences is that a limited system or set of rules can create complexity that cannot be understood by a analysis of individual elements. This idea is augmented with the notion that, in the social reality of open systems, the origins of complexity are the conjunctions of local conditions and generic patterns. In other words: social reality concerns situated complexity and this is the ontological point of departure for this book, on which we will expand in Chapter 6.

1.2.4

WHAT CAN WE SAY ABOUT COMPLEXITY?

Given that complexity is intensified by the relative position of the human observer, the question arises as to what extent we as observers are able to say something meaningful about complexity. What do we communicate when exchanging information and insights about social complexity? Researcher draws a principal distinction between epistemic, ontological and functional complexity, with each category having several subcategories of its own (Rescher, 1998). The epistemic mode concerns the complexity that is necessary to describe and analyze complexity, such as the descriptive complexity discussed in Section 1.2.1. The ontological mode concerns the types of elements and the way they are related to each other. Using a series of characters to represent elements and their mutual relationships assumes an ontological mode of complexity. The third form, functional complexity, describes the functions and workings of the elements and their mutual relationships. In other words, it concerns the dynamics and directions of complexity.

Human observers are part of the social world we observe. As such, we can't step outside social reality - society it is not an exogenous thing to us. Consequently, all three modes of complexity apply to the issues discussed in this book and this makes things even more difficult and provides little help in structuring the things we observe. Adding to the difficulty is the fact that complexity sciences consist of a myriad of notions from various origins, some of which have similar meanings under different headers or vice versa. Moreover, much of the vocabulary of the complexity sciences is rooted in natural sciences such as physics and chemistry. The notions may be commonly used in the natural sciences, but may appear exotic when applied in the social sciences. In addition, there is no natural hierarchy or order that determines which notions should be discussed first. Indeed, it is inherent to complex complexity that there is no real or obvious starting point.

With regard to the first two points, the question arises as to what the added value is of transferring concepts from one scientific domain to another. Theory transfer is not uncommon but is often criticized. Some of the criticism related to the application of foreign theories into the analysis of social phenomenon such as public decision-making is discussed in this book to allow for an open assessment of the added value of these elements. Some, such as Rosenhead (1998) and Kerr (2002), criticize authors who have somewhat carelessly copied notions of complexity from one realm to the other without adequate consideration of the assumptions underlying these notions. For example, the fact that programmed entities are able to self-organize by very simple rules of behaviour as programmed in Reynolds' computational simulation does not mean that organizations of people are always able to self-organize by similar simple rules of behaviour. Similarly, the fact that Bénard cells emerge in heated fluids does not mean that societies change according to the same principle. In short: the causation present in the source domain does not mean that the same causation is preserved in the target domain.

Underlying this critique is the ideal of conceptual purity. However, conceptual purity cannot always be maintained in a set of theories under development. Restrictive use of concepts could cut off the potential of added explanatory power and may, in turn, frustrate further theoretical development. Cilliers (1998) and Williams (2000, in Haynes, 2003) posits that while purity itself is a good thing, it should not restrict further development within the realm of the social sciences. Both Byrne (1998) and Haynes (2003) argue that the analysis of social complexity can adopt concepts from the physical sciences but that this process cannot just be pure replication as the meaning of a concept evolves when confronted with social reality. This evolution should be allowed to take place.

Chettiparamb (2006) proposes an alternative way of looking at theory transfer and theory evolution in complexity, which is to understand it in terms of metaphors. Thinking and communicating about the complexity of public-decision making in terms of metaphors can help to shed light on it (e.g. Gulick, 1984). From this point of departure, metaphors are understood as the vehicles for transfer from one field of science to another. It is accepted that the properties of the target domain do not necessarily correspond to the properties of the source and thus, that the metaphor has a dynamic meaning because of this difference. Metaphors may be used to develop analogies between the two domains and can work to enrich both, as the transformation of the metaphor during the confrontation with the target domain can help to enhance understanding in the source domain (Chettiparamb, 2006). This approach replaces the one-sided perspective that informs conceptual purity by focusing instead on the interaction between domains (Lakoff & Johnson, 2003).

An extensive survey of the literature conducted by Maguire and McKelvey (1999) similarly points out that the use of concepts from the complexity sciences to interpret social developments often relies on the functions of metaphors. However, the authors note that these metaphors are constructed without much mapping of the source and the alterations that occur during transplantation provide a very superficial understanding of the subject, thus giving very little added value. While Chettiparamb argues that there is a role for metaphors in shaping a theory in an alien domain, other authors believe that the careless application of metaphors ultimately harms the development of complexity-informed research in public administration and public policy (Mathews, Michael, White & Long, 1999). Metaphors can be used to persuade an audience to look at something in a coherent or different way but if a closer look reveals nothing but more metaphors, initial enthusiasm may turn into cynicism.

In order for a theory to gain authority, it should be able to pass the level of the metaphor and have explanatory power (Rosenhead, 1998). Rosenhead concludes that much of the initial work on the combination of complexity sciences and social sciences, especially where it concerns management sciences, barely passes that level and that the empirical foundations are either of anecdotal character or derived from the natural sciences without much consideration. This further erodes the use of metaphors as a method of theory development. He also points out that some of the oft-quoted researchers from the natural science domain such as Kauffmann and Krugman do not state that they have evidence for every argument they make. However, reference to these works as solid proof of assertion has occurred in the (popular) scientific literature. This weakens the case for complexity thinking in the social sciences.

Although the criticisms regarding the thoughtless application of metaphors is targeted at the early attempts at theoretical development in the social sciences and more elaborate accounts have been published since, and although the complexity program in the social sciences is still in its infancy, it is a clear indication that the transfer of concepts from other domains to this domain requires thorough concept mapping, either when used as a transfer by means of metaphors or when used as an explanation for certain events transcending the level of metaphors. The concepts and theories presented in this book are meant to enhance explanatory power, i.e. they are meant to increase our understanding of the complexity of public decision-making. It is for this reason that we will spend some time discussing the origins of ideas and the alterations necessary before applying it to the core subjects of the book.

The following choice is made in this book with regard to the third point on the lack of hierarchy or order in structuring the concepts of complexity theory. We will follow Cilliers (1998; 2001) and will structure this diversity into the categories of structures and processes, whilst at the same time constantly reminding ourselves that structures and processes do not exist independently from each other. Complex systems and the activities taking place in them need each other to exist. Any attempt to structure complex complexity in a way to make it comprehensible does some injustice to social complexity, but this is necessary simply in order not to get lost (Cilliers, 2001). In this book, we are aware that we could have chosen a different point of departure or a different structure of presenting the concepts and theories. Nonetheless, we appreciate that one has to start somewhere and the order of the chapters reflects a reasoned choice of structuring.

1.3 BLACK BOXES OF CAUSALITY

We started this chapter with an example of how a seemingly unrelated and random fire in a tunnel in Switzerland, severely impacted decisions about an unconnected project in the Netherlands. We made the case that such events are not deviations from the norm, but in fact, rather common. We also made it clear from the beginning our commitment to accepting all the imperfections of the world, instead of dwelling on how things could or should be. With such a stance, we seek to uncover the basic building blocks of the complexity of social reality. But how would this help us understand the issues presented in the introduction to this chapter?

Public decision-making takes place essentially because a number of people aim to change a situation that they deemed unfavorable into a more favorable one without actually being in any position to individually impose their decision on the collective. Every policy decision assumes a causal relationship between the steering incentive and its possible consequences. In other words: if people think that a certain policy *x* influences a certain situation *y*, they also assume that there is a set of mechanisms underlying this policy that makes situation *y* happen. However, it is often not self-evident which factors promote or prevent situation *y* from occurring .

Policies regarding urban regeneration serve as pertinent examples of such complexity (see e.g. Doak & Karadimitriou, 2006). Suburbanization in both Europe and the United States has brought about a shift in the spatial urban pattern, in some cases leading to the development of the so-called edge-city (Garreau, 1991; Bontje & Burdack, 2005) where new suburban cores take over the functions that were previously concentrated in the inner-city (Bingham & Kimble, 1995; Gospodini, 2006). This competition between the inner-city and the new polycentric urban networks has altered the urban landscape to such an extent that in many instances, the inner-city has been left with fewer vital functions. This stripping down of its role as a commercial and cultural centre makes the inner city less attractive in general, lowers rents and market prices, and homogenizes the population towards people with low income and purchasing power. Upon closer inspection, the relationship between the emergence of the suburban cores and the decline of the inner-city shows many interlocking factors that are mutually reinforcing, and that foster a vicious cycle (Byrne, 2001). For example, household displacement leads to the disappearance of more commercial or cultural functions and services, which in makes the neighborhood more unattractive neighborhoods and lowers rents. Lower rents lower the incentives to maintain buildings and to develop new projects, causing already run-down areas to become increasingly unattractive for entrepreneurs, and pushing households with sufficient means to relocate elsewhere. This cycle reinforces.

Internationally, there have been many varied policy attempts to reverse such cycles (e.g. Vella & Morad, 2011;

Swyngedouw, Moulaert & Rodriguez, 2002). Each such attempt to regenerate the inner-city rests on the assumption that the factors underlying the downward spiral of that area are being addressed. However, the outcomes of such efforts for the whole cycle are far from certain because changing one or a number of factors may also bring about different effects. In fact, addressing the cycle incorrectly could reinforce its unfavorable outcomes. Findings from e.g. Cameron and Doling (1994), Swyngedouw, Moulaert & Rodriquez (2002) and Atkinson (2004) show how policies aimed at the gentrification of the inner-city can lead to a range of unintended effects, including a reinforcement of the vicious cycle. In such cases, the effects of the policy bring about the exact opposite of what it aimed to achieve. Clearly, urban regeneration policies need to carefully address connected factors in order to diminish the chance of bringing about unintended effects. In other words: urban regeneration takes the form of a *systemic* set of activities (Atkinson, 2004) but without the illusion of total control over the urban system such as the failed attempts at blue-print urban planning (e.g. Smets & Salman, 2008). And since every city has its own particular conditions, a recipe that works in one place might not work in another (Pierre, 2005).

Developing and deciding upon sound policies requires an understanding of both the intricate workings of social reality and the way public decision-makers and their administrations operate, as both the individuals and their organizations are integral parts of a system of causes and consequences that shape change. Knowledge of the workings of the social and physical system is available in scientific domains such as sociology, economics, human geography and environmental studies. Such knowledge may shed greater light on the possible uncertainties, restraints and outcomes of decision-making. However, public decision-making is still generally regarded as a black box from the perspective of such domains as little is known about its dynamics and its impact on societal concerns. Thus, is can happen that some authors arrive at advanced analyses of complex issues and still recommend policies that are built on simplistic premises. The reverse goes for the domain of Public Administration and related studies. While the dynamics of public decision-making are its core subject, less is known about the effects of decisions and how these effects influence decision-makers and decision-making at a later stage.

In the domain of Public Administration it is the environment that is the black box (Parsons, 1995). The complexity of public decision-making can only be better understood if the two black boxes are opened and connected. In other words: to answer the question of why much of public decision-making is complex, it is necessary to understand the mechanisms underlying environmental complexity and its reciprocal relationship with the acts of public decision-making.

There are a number of different methods of inquiry that will shed light on this question. A considerable body of work has accumulated that is based on experiments or advanced modeling (e.g. Otter, 2000; Koliba, Zia and Lee, 2011). These attempts are valuable in mapping the complexities of the patterns that emerge based on pre-set assumptions. However, we are more specifically interested in this book in the inherently messy, day-to-day dynamic practices of public decision-makers. Although any attempt to answer the main question should be fueled by in-depth empirical research, given the messy reality of decision-making, we can't be certain that we've seen all there is to be seen. In researching complexity, one often finds that uncovering one mechanism reveals another set of related mechanisms underlying it. Our aim in the next five chapters will be to refine our attempts.

1.4 OVERVIEW OF THE BOOK

As mentioned before, there is no natural or given order that determines which notions should be presented first. For example, it is not possible to see complex systems detached from the processes that propel those systems. However, not everything can be presented at once so choices must be made about which factors to address first. While the chapters may be read in random order (if so desired), the argument does advance throughout the book and it would help to read the chapters in the way they are presented. In short, the book discusses the systemic (Chapter 2) and dynamic (Chapter 3) nature of public decision-making, how humans cope with complexity (Chapter 4), how they arrive at decisions (Chapter 5), and how this complex whole can be effectively researched (Chapter 6).

More specifically, Chapter 2, ('Being Dependent') focuses on structure and introduces the subject of complex systems and the intellectual development that systems' thinking has been through. It shows what systems are, how they exist in reality and why they matter for public decision-making. The boundaries of systems, i.e. what separates a system from its environment, are subject to debate since the act of drawing boundaries has important consequences for the outcomes, as can be surmised from the examples in this chapter. Chapter 3, ('Being Dynamic') builds on the systemic premises discussed in Chapter 2 and focuses on the processes that make systems tick. It presents a number of notions such as the positive and negative feedback, hysteresis and path-dependency that helps explain both inertia and dynamics as a property of systems, and how well-intended policies can bring about unexpected results. In Chapter 4 ('Being Human'), the focus is shifted to human behaviour and its role in the complex systems as presented in the previous two chapters. It discusses specifically how people cope with the complexity of their environment when attempting to make decisions. It demonstrates how complexity impacts the human mind and how it drives the choices that are made. Inevitably, it is part of the human condition to simplify complexity, and it becomes necessary to look at how this simplification operates and impacts decision making. Chapter 5 ('Being Coevolutionary') gathers the strands from the previous chapters and combines them to present a coevolutionary approach to understanding the public decision-making processes. The main feature of this approach is a shift away from decision-makers as the loci of change towards a systemic view of change or lack thereof. Thus, the concept of coevolution helps bringing different ideas about public decision-making together. The sixth and final chapter ('Being Inquisitive') lays out the ontological and epistemological foundations of the book's approach to complexity. It proposes a complexity-informed research approach to analyzing public decision-making. The main motive for putting this chapter at the end of the book is so that it won't distract readers who are more interested in the implications of complexity than in a scientific debate of how to investigate it. Still, it is an important chapter as it presents the foundations for the statements made in this book. And now, without much further ado, let's talk about complex systems.

CHAPTER 2.

BEING DEPENDENT: COMPLEX ADAPTIVE SYSTEMS.

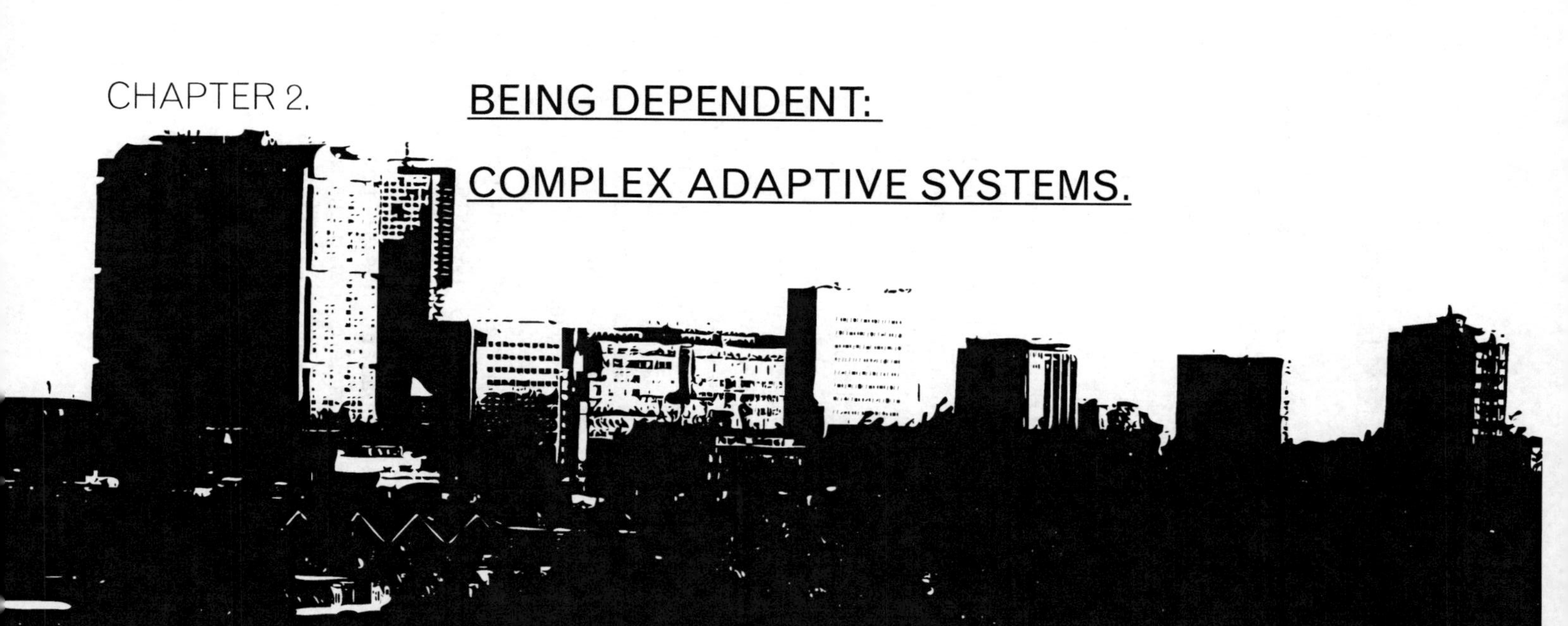

2.1
INTRODUCTION

The previous chapter established the roots of complexity and gave a tentative answer as to how complexity emerges and grows. In this chapter, we address the idea that complexity is a property of systems, a notion so commonplace that it is often not mentioned or explained (Phelan, 1999). The term 'system' itself is also used in science and in everyday parlance as liberally as 'complexity'. The word 'system' hints at the existence of multiple relationships between different elements. The idea that phenomena were systematic and operated as such dates back to the 17th century (François, 1999). Many systemic theories and different schools of thought have emerged since then, and some of them still exist and continue to evolve (Waldman, 2007; Cabrera, Colosi & Lobdell, 2008). For example, a clear distinction has emerged between systems thinking and systems theory.

Systems thinking has seen multiple revivals in scientific literature and practice and has inspired many efforts to solve real-world issues. However, what systems thinking precisely means remains somewhat elusive. It refers to "a different way of thinking and seeing" (Meadows, 2008: 4), and "provokes thinking about the world in a completely different manner than other forms of thinking" (Houghton, 2008: 100). Others have said that it can "influence many of the existing concepts, theories and knowledge" (Cabrera, Colosi & Lobdell, 2008: 299) because it challenges "simplifying assumptions" (Merali & Allen, 2011) and provides a "major alternative to the reductionist and discipline-bound mainstream in the social sciences" (Barton, Emery, Flood, Selsky & Wolstenholme, 2004: 4). Systems thinking also helps to "focus on processes, interactions and causes of poor outcomes rather than individual players, isolated components of a system or interim results" (Waldman, 2007: 279). If the many authors writing about this topic have one thing in common, it is the view that reductionist explanations are not accurate representations of the world, but rather are accurate representations of our cognitive limits in understanding the world around us. It is tempting to search for that one variable that explains outcomes in each instance, but a systemic worldview holds that each result is an outcome of a particular situation that is local in place and time. This is an issue which we will return to frequently in this book. Perhaps most significantly, systems thinking offers a worldview with a specific approach towards causality and consequences that changes the way people deal with this world. Flood (1999) asserts that systems thinking "takes issues with (the) grand narratives of strategic planners who think globally and believe that with intention they can create a better future. In their reports they innocently indulge in fictional script writing."

As the distinction between systems theory and systems thinking suggests, thinking in terms of systems is not the same as coming up with a theory of systems. A theory is a set of causal mechanisms rather than an elegant heuristic concept. The goal of this chapter is to provide the reader with an overview of the way we have historically thought about systems, the issues that arise with each different version of system theories, and the implications this has for public decision-making. Consequently, it has a much broader and more introductory scope than the other chapters in this book. In particular, it explains why public decision-making is systemic and how systems contribute to our understanding of the complexity of public decision-making processes. A brief overview of systems thinking and systems theory is first presented (Sections 2.2.1 and 2.2.2), followed by an introduction to the importance of meaning and learning (Section 2.2.3). We will then discuss the development and application of systems within the realm of public decision-making (Section 2.3.1) and the relationships between systems, boundaries and hierarchies in decision-making (Section 2.3.2). The concept of complex adaptive systems is then introduced (Section 2.4.1) to better understand the systemic and complex nature of public decision-making. Research into the operation of the Dutch railway system is used to demonstrate the workings of such systems (Section 2.4.2). Finally, homogeneous and heterogeneous policy action systems are distinguished from complex adaptive systems to shed light on the nature of their adaptive moves and the layered character of complex adaptive systems (Section 2.4.3).

2.2.1
A FIRST EXERCISE IN SYSTEMS THINKING

The use of the word 'system' is relatively common in both everyday parlance and science, but how does one know when one is looking at a system? As Cabrera, Colosi & Lobdell put it, one sees the whole first and its components later (2008). That is still somewhat abstract, so it is instructive to follow Meadows (2008), who identifies three basic conditions for systems: [1] elements, [2] relationships between elements, [3] a function or a purpose. She presents these elements as the basic building blocks. The function or purpose is the condition that keeps the system intact because it allows individual elements to be replaced without the whole system collapsing. These conditions do not describe a particular system, and observers can further define and operationalize a system.

Meadows' thought experiment serves as a first approximation of systems thinking and demonstrates that defining systems is not as straightforward as it may first seem. For example, what are the tangible and intangible traces of relationships? More importantly, how does one assign a function or purpose to a system? Meadows and many other authors acknowledge that system definitions are elusive in the sense that each definition often leads to further refinement. Defining systems is a continuous process that may start with the three conditions mentioned above, but then needs to advance further into a more refined definition. It is this kind of defining and redefining that haunts systems thinking.

Systems can be further understood using the analogy of stocks and flows (e.g. Sterman, 2000). Stocks are "the elements of the system that you can see, feel, count, or measure at any given time" (Meadows, 2008: 17). Flows express the movement of stocks. A steam engine may serve as a mechanical example here. If water in the engine boiler is a stock, the heat turning the water into steam, the steam itself, and the replenishment of the water and fuel are all flows connected to that particular stock. It is possible to manipulate the flows. For example, the heat can be increased inside the boiler such that the resulting steam pressure is regulated. Thinking in terms of stocks and flows improves our understanding of systems as diverse as that of engines, factories and entire markets.

The stocks and flow systems discussed in this book have people at their center. Obviously, people exerting their influence are an integral part of public decision-making systems. Such individuals have normative ideas about what they want and how they want it. They are also reflexive and able to adjust their roles and positions according to these normative ideas. They anticipate, plan and act deliberately. They also attempt to make forecasts by guessing what might happen next and what they should do to influence outcomes. In other words, humans play an active role in the operation of systems, including public decision-making systems.

The involvement of thinking humans who can act in different ways creates "thinking systems" (Waldman, 2007). The ability of people to respond differently to situations, even those that they have experienced in the same way before marks the creativity that is present in such systems. To us, it means that systems can and will surprise us, even if we have made strenuous efforts to understand their nature and operations. The inclusion of humans in systems also underscores the fact that the distinction that is often made between human and non-human systems is entirely fictional. As argued in actor-network theory, humans use physical objects for their own aims, but the physical objects themselves influence humans (Latour, 2005). Systems can therefore be seen as combinations of human and non-human elements and their mutual interaction.

An example of how human and non-human elements interact and how this interaction influences the operation of systems can be found in Rotterdam, where the local authorities found the groups of teenagers that loitered around the Zuidplein metro station and the mall next to it to be a nuisance. Using the police to chase them away did not help much. While the police dispersed the groups temporarily, they returned over time and these encounters established the group's street credibility. Understanding this, the municipality installed a barrel organ at the metro station and mall which played old-fashioned Dutch music. The teenagers found this music to be decidedly un-cool and not surprisingly, they subsequently chose to stay away. How teenage groups become established and gain credibility can thus be seen as a system that the municipality intervened in. They did so by introducing a non-human element which subsequently became part of the policing system.

BIJ OVERLAST MOSQUITO ACTIEF
Waar staat mijn bus op Zuidplein

Many thinkers, in both the public and business administration fields have written extensively about issues related to problem definition and mapping. In the following section, we review their ideas.

2.2.2 THE ORIGIN OF SYSTEMS

François (1999) explored the emergence of systems thinking in the fields of philosophy and science and found the earliest mentions to be in the 17th century works of Descartes and Leibnitz. These philosophers argued that philosophy should be systemic, i.e. that it should be built on a coherent set of postulates that leads to a systemic philosophical construct. It is sometimes hard to pinpoint whether a certain thinker aimed to develop a systemic framework, or whether the systemic framework resulted from certain logical steps. Regardless, all contemporary systems thinkers are indebted to Ludwig von Bertalanffy, an Austrian-born biologist, for furthering this line of thought. According to Flood (1999), von Bertalanffy (1901-1972) was one of the first modern systems thinkers, and he operated in a time when reductionism was the most common method of understanding the physical world. Reductionism has as its premise the idea that the operations of a certain phenomenon can be induced from the workings of individual parts. It seeks out one or a few main factors that explain the variance in the phenomenon. Reductionism involves the elimination of environmental variables thought to be irrelevant to the problem in order to improve explanatory power. However, von Bertalanffy spoke loudly of the limits of reductionism in 1920s. "Von Bertalanffy demonstrated that concepts of physics were helpless in appreciating dynamics of organisms. The existence of an organism cannot be understood solely in terms of the behavior of some fundamental parts..." (Flood, 1999: 29) He showed that the nature of organisms was partly determined by their relationship with the environment, and argued that this needs to be understood to explain the nature of the system (Von Bertalanffy, 1968).

He applied this idea to other non-biological examples and is credited with coining a general systems theory that (by being systemic) purports to identify the main drivers of all types of systems. He fully appreciated that different systems in different domains have different properties. Nevertheless, he posited that:

> "... there exist models, principles, and laws that apply to generalized systems or the subclasses, irrespective of their particular kind, the nature of their component elements and the relations or "forces" between them. It seems legitimate to ask for a theory, not of systems of a more or less special kind, but of universal principles applying to systems in general. In this way we postulate a new discipline called *General Systems Theory*. Its subject matter is the formulation and derivation of those principles which are valid for "systems" in general." (Von Bertalanffy, 1968: 32; italics original)

Two ideas are central to his theory: first, that there is a continuous exchange of energy and information between organisms and the environment, and second, that it is necessary to accept that systems are open in order to understand the way systems behave. His theory of open systems has a compelling logic but also poses two major issues. The first issue is that thinking in terms of open systems potentially means that an unlimited number of variables have to be taken into account to understand phenomena. Nonetheless, open systems theory has an advantage over closed systems theory and reductionism because it emphasizes that systems are partly determined by their environment. In more abstract terms, systems are embedded in other systems, which are in turn embedded in yet other systems (Byrne, 2005). For example: an increase in the number of migrants in a certain neighborhood can be traced back both to a flood in another village at the other side of the world and to climate change, as well as to the presence of an actively employing company in that neighborhood. While these causes are plausible, it is often difficult to conclusively determine the exact causes. Thus, the advantages of open systems theory must be considered together with the increased uncertainty introduced by the greater number of variables that are taken into account.

The potentially unlimited connectedness of systems leads to the second issue. It makes it necessary to demarcate the system boundaries to keep research into them, and possibly efforts at steering them, manageable. Because this decision has to be made, what defines and limits a system is subject to negotiation (Cilliers, 2005a). The issue is complicated by the fact that systems are often intangible or ambiguous. This challenge is connected to semiotics, which is the process of assigning meaning to the world. For all the merits of his ideas, von Bertalanffy did not fully account for the fact that definitions of systems are projections of people's values, ideas and beliefs, and that these significantly affect the demarcation of the parts of systems that they deem relevant. Is it possible to draw an uncontested boundary around the neighborhood discussed above? A geographical boundary, marked by streets and other such elements, would not draw much debate. But a geographical boundary does not stop the flows into and out of the neighborhood described above. Burgers (2002) therefore redefined such neighborhoods as 'landing-stages' in a much wider system of migration, analogous to the arrival gates found in airports.

Von Bertalanffy's legacy can be identified in most contemporary accounts of complex systems, regardless of whether they are about social systems or other types of systems. For example, the environmental sciences have long conceptualized ecologies as complex systems (e.g. Hartvigsen, Kinzig & Peterson, 1998; Levin, 1998; 1999; Certomà, 2006). Gerrits (2008) is an example of how researchers in this domain think about systems and how this affects public decision-making. This is a case study of the deepening operations of the Westerschelde estuary, which runs between the North Sea and the port of Antwerpen. The Westerschelde is one of the last major estuaries on the European west coast and many consider it a valuable ecological habitat. It also provides maritime access to the port of Antwerpen. Because of its limited depth, the Antwerpen port authorities want to make it deeper so that bigger ships can reach the port around the clock. However, since the Westerschelde is also an important ecological system, one of the main issues facing public decision-makers was the environmental impact of deepening the estuary.

The researchers defined the estuary as a complex system consisting of multiple channels, shoals and sand bars. These different elements kept each other in an unstable equilibrium and much of the policy research focused on estimating the probability that this system would topple if the estuary was deepened. The factors that were taken into account included the relationship between sediment transport and estuarine biology, and the exchange of sand between the estuary and the North Sea. More variables were considered so as to better understand the complex processes that characterized the estuary; however, this practice increased exponentially the level of uncertainty over the possible outcomes. Clear-cut answers could not be given, and the lack of decisive information was used as argument both to speed up the decision-making process and the deepening work ("we won't know if we don't try it") and to stall it ("more research is necessary before decision is made").

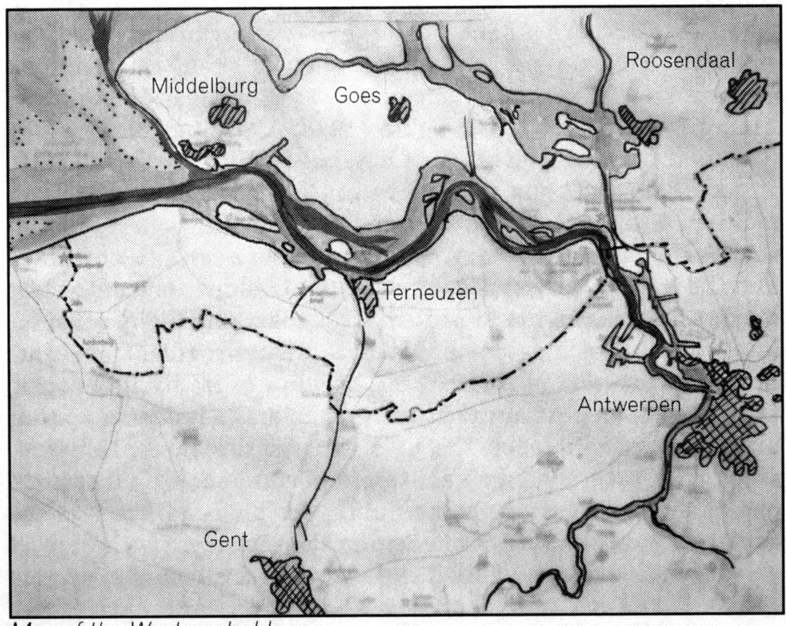

Map of the Westerschelde

HONTE

A focus on the complexity of systems is also present in some sub-domains of economics with traces of biological and ecological thinking as foreseen by von Bertalanffy (e.g. Arthur & Durlauf, 1997; Smith, 2002; Kirman, 2004; Foster & Hölzl, 2004; Earl & Wakeley, 2010). Understanding economics as complex systems means reversing many of the assumptions that underlie neo-classical economics, such as a shift from rational individual agents to an erratic and heterogeneous population whose decisions influence the state of the systems (Bergh & Gowdy, 2000). This change incorporates the complexity of the systems from which choices regarding future options are made into its analysis (Foster & Hölzl, 2004). An example of this is presented by Frenken and Nuvolari (2004) who investigated the evolution of propulsion technologies, such as aircraft and helicopter propulsion. They proposed that competing propulsion technologies could co-exist without one of them becoming dominant over time. This view contradicted the commonly-held belief that technological evolution was a linear progression where better technologies substitute lesser ones. They demonstrated that a systemic set of interdependent technologies that performed various functions was more likely to exist. Technological evolution therefore occurs in the form of combinations of technologies rather than single technologies isolated from the system they are embedded in. In this view, technology design is therefore not the matter of designing the 'best' technology but rather "assembling the right set of components in a functioning system." (2004: 97)

The number of possible combinations is very high and designers are likely to explore only a limited set of designs instead of exhaustively trying all combinations. In this way, design becomes a local search process in the system of interlocking technologies. This systemic redefinition of technological change explains the occurrence of multiple co-existing clusters of aircraft and helicopter propulsion technology. The different clusters find their own niches in the aircraft and helicopter market where the combination of the technologies matters more than the performance of individual technologies.

Such systemic approaches have also emerged in the social sciences. An important contribution came from Talcott Parsons, an American sociologist and biologist, most prominently in the book 'The Social System' (1951). It presented a socio-cultural approach to understanding societies, without the conceitedness of a holistic explanation of all things social. It proposed a unified theoretical framework for studying social phenomena (Checkland, 1981) that distinguishes between social, physical and cultural objects that together form systems. As such, the physical and social realms become connected, and the artificiality of the demarcations between the two becomes starkly apparent. He approached actors, both as individuals or as groups of individuals, as purposeful, functionalist elements of the system.

Parsons' work on social systems received both favorable and unfavorable reviews (Turner, 2005). The criticisms that are relevant for this book are twofold. First, Parsons' conceptualization of systems puts heavy emphasis on the stability of systems, i.e. he defined the normal state of systems as being homeostatic, with systemic changes being deviations from the norm. The stability of the system and its assumed tendency to return to the same equilibrium state over time was attributed to the functional contributions of the system's elements to the system as a whole. Parsons' functionalist worldview "appears to be one of conservatism and conformism, defending the system [...] as is, conceptually neglecting and hence obstructing social change." (von Bertalanffy, 1968: 196) This criticism is rooted in the observation that social systems could display a certain degree of stability but that this stability is matched in equal amounts with instability and change. If social systems are in a state of equilibrium, it is an unstable equilibrium.

Second, Parsons' work "incited the aspiration to develop all-inclusive models of society for the purpose of policy-making in different sectors of society, and to frame theories according to those models. Ideas of a systems approach were combined with ideas on rational decision-making to generate theories that stressed a comprehensive understanding of society with a rational weighing of alternatives." (Klijn & Snellen, 2009: 18) Here, the structural-functionalist worldview coincided with rational decision-making by purposeful actors, motivated by optimization. It heralded the arrival of total systems thinking in which this systemic worldview was combined with ideas about how societies ought to be steered or governed. Parsons' work led to the adoption of systemic approaches to societal steering such as comprehensive planning procedures and the establishment of planning bureaus (Klijn & Snellen, 2009).

Edward Quade, a mathematician and physicist who worked for the RAND Corporation, can be credited for integrating operations research with a functionalist systemic approach in a method that he called (rather broadly) 'policy analysis' (1975). He attempted to comprehensively analyze the consequences of policies to advice public decision-makers as to the 'correct' course of action. In short: "systems or policy analysis [...] is a form of applied research designed to acquire a better understanding of sociotechnical problems and to bring about better solutions" (Majone & Quade, 1980: v). Originally, Quade's work concerned defense and engineering but later also extended to less technical or 'softer' systems (Checkland, 1981). His policy analysis is a mix of Parson's functionalism and systems engineering, and he derived a model of synoptic public decision-making that spanned from problem formulation through to policy implementation, incorporating post-implementation feedback. In this case, the implications of systems thinking were used to develop an approach toward decision-making that is firmly rooted in systemic thinking.

Another explicit attempt to combine a social system's approach with the logic of public decision-making was made by David Easton, a political scientist. His use of systems theory to model the processes of public decision-making is of particular interest here because "... it provided a model of the 'political system' which greatly influenced the way in which the emerging study of policy (outputs) in the 1960s began to conceptualize the relationship between policy-making, policy outputs and its wider 'environment'" (Parsons, 1995: 23-24). The model contextualizes the act of public decision-making by showing how it is affected by the societal environment and how the decisions made within the policy system in turn affect society through feedback loops. The societal environment includes, according to Easton, social systems, ecological systems, personality systems and political systems elsewhere. This makes Easton's model a prototype for thinking about public decision-making and public policy as systemic activities.

One of the main characteristics of this school of thought is an inclination to view society as a whole comprising of parts, similar to a watch that can be manipulated to control it. The search for the one parameter that controls society turned out to be futile, and this led to disappointment with the systemic approaches to policy and public decision-making (Otter, 2000). The reason for this failure is two-fold. Firstly, societies do not operate like mechanical watches. Causal relationships between different elements are not necessarily permanent. Secondly, each causal relationship is contingent, i.e. it depends on the local situation. As discussed in the previous chapter: what may work in a certain situation may not work at all when repeated in another time or place. Thus, the scientific quest to develop the 'correct' causal recipe to tackle society's problems was destined for failure.

2.2.3 INTRODUCING HUMANS, AND THE EFFECT OF SEMIOTICS AND LEARNING

The development of systems thinking and systems theory is one of hope and optimism alternating with disappointment and criticism. This is especially so in the realm of public administration. The association of the word 'system' with holistic and rigid steering concepts is enough cause for some to dismiss it altogether. However, the value of systems concepts lies in their contrast with mechanistic approaches. Certain conceptualizations of systems accept that reflexive humans are an integral part of the systems. Therefore, "understanding does not simply rise from observation or theory" (Flood, 1999: 55) because no one can be a neutral outsider of systems. Thus, semiotics is important in the definition and operation of systems. There are many accounts of complex systems where this aspect is not addressed explicitly, resulting in confusion about whether a physical system or a social system is being described (Rosenhead, 1998). The cascade of literature on systems is therefore characterized by the (re-)discovery of the highly varied and diverse systems that are populated by learning, reflexive and adaptive humans and of which the boundaries are not given but subject to negotiation.

Despite recurring disappointment, systems thinking has been ongoing in many disciplines. Flood's (1999) rich account shows how far thinking and theorizing about social systems has advanced in the realm of decision-making in the private and public sectors. Apart from Flood, important researchers in this field include Checkland, Churchman, Ulrich and Senge. These writers put semiotics and learning at the core of their system theories and argue that while mechanistic or 'hard' system analysis has its merits, it ignores or downplays the systems of meaning that are crucial to social beings.

To Checkland, systems shouldn't be researched as if they are tangible things 'out there'. Instead, they should be understood through the researcher's action. His point of departure is that a functionalist definition of systems is misguided because it only manages to capture the tangible traces of systems. It leaves aside the intangible or 'soft' dimension of semiotics, even though semiotics is what makes a system. No wonder then, Checkland reasons, that systems engineering works with engineered systems such as manufacturing and transportation logistics, but fails to deliver with systems that focus on humans. He came to the conclusion also that the then-popular image of the researcher or analyst positioned outside the system, analyzing with it and tinkering with it, was understandably appealing but analytically incorrect. It may apply to a factory engineer who tries to optimize the manufacturing process, but it does not apply to human or social systems (Checkland, 1981).

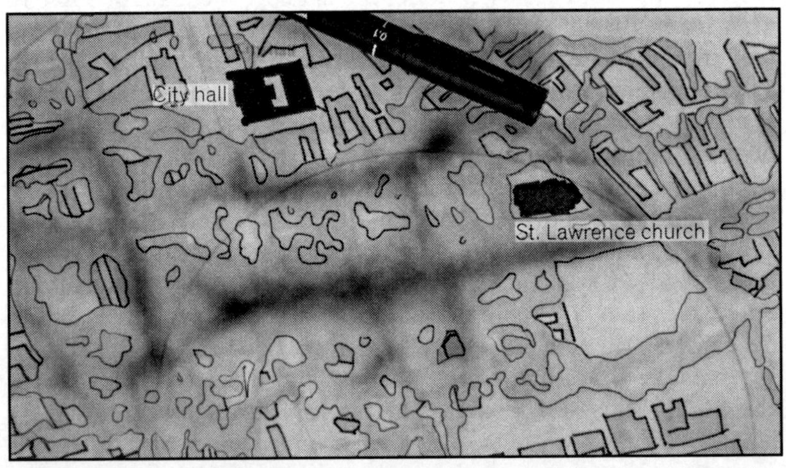

Cities are systems with tangible elements (e.g. buildings) and intangible elements (e.g. human behavior). This map shows the walking patterns of people, equipped with GPS-trackers, living in high-rise buildings in Rotterdam.

This conception led Checkland to adopt a radical position concerning how systems should be analyzed: he discards the difference between researchers and practitioners because neither can step out of the system. According to his view, a researcher should learn the system by working in it rather than observing it from the outside. His idea of system analysis consists of loops between action and subsequent reflection. Checkland also places great emphasis on the fact that pure, self-evident, observation does not exist and that each observation requires the development of a particular explanation to become a meaningful observation. By definition, such an explanation is colored by the observers' points of view. By combining action and reflection, the process of developing meaning becomes part of both observation and action. Thus, if one adopts Checkland's position, it becomes inevitable that a system definition turns into a system methodology. After all, one can only learn the system through action-and-reflection, a fixed pair of methodological activities: "The fact that the methodology is a learning system, and, in tackling unstructured problems *could only be* a learning system [...] is due to the special nature of human activity systems." (1981: 214; italics original) Checkland thus promotes sense-making from a supposedly bias-introducing factor and brings it to the core of system analysis.

Note that Checkland adopts the old Thomas theorem: if men define situations as real, they are real in their consequences. To him, action in systems is driven by meaning, because the latter is related to the perceptions of systems and their interpretation by individuals operating in them. In this way, systems become real through meaning and consequent action, instead of through objective or functionalist facts. This approach to systems abandons the ideas that systems can be understood independently of individual perceptions and operated independently of individual judgments. Instead, they are based on experiential learning about what the right courses of action could be. The focus on interpretation, meaning and insight through action led Checkland to dub his approach 'soft systems methodology'.

Churchman and Ulrich focus on how systems emerge through individual judgments, especially in the case of boundary judgments, i.e. the demarcations between a system and its environment. Such judgments set the problem-and-solution space of an issue, and altering those boundaries enhances the understanding of the systemic nature of the issue. Flood stresses that this approach does not lead to an absolute overview of the full system in the long run, but rather "helps people think or debate their way out of mind traps [...] or mental models [...]" (Flood, 1999: 65). This approach counters two obstacles in systems thinking: a full overview of the system is close to impossible, and trying to obtain such an overview would require so much effort that it would never get done. Senge (1990) also takes into account the mental models of reality that individual build and pays attention to the learning that takes place among individuals and groups of individuals. Learning comes from interaction and dialogue, especially if one questions the assumptions underlying the mental models. It may cause mental models to shift, reframing them in terms of Schön & Rein (1994; Rein & Schön, 1996) and in turn leading people to adapt themselves to ever-changing contingencies. It is therefore necessary to focus on setting boundaries in public decision-making and on the assumptions underlying that act, instead of trying to develop unambiguous system boundaries. This thesis is in fact a reflection of a larger issue with social systems, namely that they are essentially socially constructed and that structure and action come together in mental representations of systems (Clark and Crossland, 1985).

Flood integrated the insights from the thinkers mentioned here with the complexity sciences, on which this book is also based, to arrive at three main conclusions about analyzing and acting within social systems: systems are essentially unmanageable, they cannot be fully organized, and it is impossible for people to fully know systems. In other words: the world does not consist of separate systems, but is by definition *fully systemic*. Note that that his argument is not nihilistic. It is a call to be modest about what can be known about systems and what can be achieved in them, and is not a call to abandon systems thinking because the alternative will lead to even less knowledge and fewer clues about how to operate.

This overview of systems thinking and systems theory may give the impression of a linear progression in science and practice. However, systems thinking and analysis has developed in so many domains that it is no surprise that many varieties of it exist. Although these varieties differ greatly in form and application, they all bear traces of the core concepts discussed above. There is evidence of a maturity in that earlier notions have been amended, altered, improved, rejected or rediscovered (Barton, Emery, Flood, Selsky & Wolstenholme, 2004). The metaphors used to describe systems have also evolved from mechanistic objects such as watches and complicated machinery toward more organistic, biological and ecological conceptions. Following from this, the idea of homeostasis has also been replaced with the idea of multiple equilibriums and unstable stability (see for example, van den Bergh & Gowdy, 2000), and the social construction of system boundaries has gained more prominence. Barton et al. (2004) summarize the group of less mechanistic system definitions as *complex adaptive systems* or CAS.

There are about as many definitions as there are authors writing about complex adaptive systems. Each domain, and perhaps even each instance in which systems thinking and systems theory are developed and applied, necessitates a tailor-made definition. There is no unified system theory that is broad enough to capture all systemic properties, and diverse enough to capture the rich detail of particular instances. This imperative also applies to this book, which means that we need to have a closer look at systems in public decision-making before defining what complex adaptive systems are in our scope. In the next section, we shift our lens towards the inner systemic workings of public decision-making before returning to the issue of complex adaptive systems.

2.3.1 SYSTEMS IN PUBLIC DECISION-MAKING

Easton's model mentioned in the previous section may be seen as a first attempt to view public decision-making in relation to its environment. The model has many merits, most importantly its explanation of the relationship between inputs, decisions, and outcomes. But it is criticised for treating the organisation of decision-making as a proverbial black box. Public decision-making takes place in multi-actor settings and concerns the weighing of different values. Regardless of whether this weighing is done through direct democracy, representative democracy or even oligarchy, there are always a actors with diverging interests who struggle to get their ideas through.

The relationships between these actors are impacted by the distribution of power across them. Some have conceptualized power as being concentrated at the top of the hierarchy of a governmental organization. However, pure hierarchies do not exist within governments. Neither do they exist between governments and the societies they govern. This is because resources such as money and knowledge are distributed across actors and systems, making power very much bound to actors and systems (Avelino, 2011; Gerrits & Meek, 2012). Power enables actors to influence decision-making, meaning that decisions are determined by the power struggles that take place between actors. The negotiations over the deepening of the Westerschelde mentioned in Section 2.2.2 are an example. The Westerschelde runs through the Netherlands but serves the port of Antwerpen in Belgium. Neither of the countries has been able to unilaterally enforce a decision on the other without retaliation. The Belgians have the funds and European jurisdiction, while the Dutch have the legal powers stemming from sovereignty. Also in the mix were local authorities with different agendas and means and powerful stakeholders, such as environmental and agriculture pressure groups. As a result, a system emerged in which the different interdependencies maintained the system's structure.

Attention for the (uneven) distribution of power between different groups and the consequent pluralist perspective emerged in the 1960s and 1970s through the work of, among others, Dahl (1958; 1961), Cobb and Elder (1972) and Lukes (1974). Klijn and Snellen (2009) described this period as one of intellectual change, where the idea of a central and somewhat omnipotent decision-maker was abandoned for a pluralist approach that recognized that decision-making takes place in multi-actor settings where power is not (only) dependent on formally assigned authority, but also on the resources that actors can possess. These actors include pressure groups, (non-) governmental organizations and private companies. Pooling resources enables actors to get things done or to obstruct others from getting things done.

Further concern over the distribution of power led to the development of policy or governance network theories in the 1990's (Rhodes, 1988; 1997; Klijn, 1996; Kickert, Klijn &

Koppenjan, 1997; O'Toole, 1997). Rhodes emphasised the self-organizing capacity of networks to exchange the different resources necessary for the future of a certain issue (Kjær, 2011). If actors, such as the Antwerpen port authorities, want to get something done, it is necessary to engage in such an exchange of resources. But while resources bring actors together in the so-called policy arenas, these actors tend to be exceedingly diverse in their beliefs, interests and perceptions. In other words, policy networks exhibit a simultaneous convergence and divergence of the actors. The position of the government, itself a loose collection of different actors, is more or less equal to others in the network, not in terms of its properties but rather in what it can achieve. The Westerschelde case shows the extent of resistance the Dutch central government faces in its decisions: it comes not only from locals, but also from pressure groups and local Dutch governments. A draconian effort was required to enforce the Westerchelde decision in 1996 because the resistance was well-organized and had the means to swing things their way.

A number of authors have noted the many similarities between policy networks, systems and complexity, such as Doak and Karadimitriou (2007); Morçöl & Wachhaus (2009) and Koliba, Meek & Zia (2010). The similarities include the emphasis on multilateral relationships, heterogeneous composition and the pluralism of power distribution (Klijn & Snellen, 2009). A common definition of governance networks is of "a set of relatively stable relationships which are of a non-hierarchical and interdependent nature, linking a variety of actors, who share common interests with regard to a policy and who exchange resources to pursue these shared interests acknowledging that cooperation is the best way to achieve common goals" (Kickert, Klijn and Koppenjan, 1997: 1). Note how it is founded on the stability of relationships, common interests and a shared understanding as to how those goals should be achieved. A systems-informed definition would add that actors are also systemically related when they do not share the same goals or visions, and would emphasise the production of meaning and system boundaries, rather than primarily focusing on the exchange of resources (Koppenjan & Klijn, 2004).

The importance of boundary judgments for understanding the systemic properties of public decision-making is clear when the organisation of government is examined. Government as a whole consists of numerous organizations that in turn consist of multiple sub-units or teams or sections: systems within systems within systems, or nested systems. It is natural to demarcate such systems by formal organisational boundaries, but as we have seen before, such demarcations can be artificial. Organizational boundaries are determined by what people feel they are part of, which could include legal considerations and an exchange of resources, but not necessarily so. Niklas Luhmann's work will help to clarify this.

2.3.2 BOUNDARIES AND HIERARCHIES IN GOVERNMENTAL ORGANIZATIONS

German sociologist and public administration scholar Niklas Luhmann developed a general theory of social systems, paying ample attention to the role of public decision-making, government and bureaucratic systems (Brans & Rossbach, 1997; Schaap, 1997a; 1997b; Seidl, 2005; Seidl & Becker, 2005). It is no surprise that as a student of Parsons, he was first heavily inspired by structural functionalism. However, unlike his teacher, Luhmann placed more emphasis on semiotics and the role of communications and expectations in establishing system boundaries (Schaap, 1997a). Like some of the researchers discussed above, Luhmann builds on the idea that people develop and maintain mental models from which they reason and act because of the overwhelming complexity of the environment. These models develop both consciously and unknowingly. How that happens is discussed in Chapter 4. What matter here is that such models help to lessen the complexity of systems by ordering them in terms of organization and structure, or to put it more precisely: "[...] the ultimate function of social systems (as defined by Luhmann – LG): the grasping and reduction of complexity" (Bednarz, 1984: 55).

From this perspective, the organization of government is essentially shaped through communicative connections between people. Bureaucratic communications are distinct from informal communications because much of the former is captured in protocols, standard operating procedures and organizational identities expressed in such protocols. However, formalized, commands and acknowledgments, norms and accountability transferred through the bureaucratic system are still communicative acts that are subject to the same dynamics of other kinds of communication. It is important to note how communicative acts enforce the system-environment dichotomy in the organization of the government (Luhmann, 1981). It is through this view that it becomes obvious why governments as organizations do not disappear when people move in or out because the identity of the system is maintained through communications. This does not mean that a bureaucratic system is a thing as such but rather that its expectations are institutionalized (cf. Williamson, 1998; Sanderson, 1990) to such an extent that they force adaptation to the existing norms. In other words: if one wants to become a member of a system, one has to act accordingly (March, 1991; 1994). This reinforces the nature of that particular system. Essential to Luhmann's theorisation of systems is that their degree of openness varies and their boundaries are subjective and negotiable. To us, it means that we have to rethink the habit of treating a group of organizations around a policy issue as synonymous to a system. It may, and it may not be, and both possibilities could occur simultaneously.

The previous sentence may sound absurd but it makes sense. If we accept that systems are open and that their boundaries are social constructs that emerge through semiotics, a person who defines systems as coinciding with formal organizations and acts accordingly has a different 'real' system than a person who uses a different definition and who acts according to his or her own definitions. Each situation therefore demands a situational approach (Fuchs, 2002; Morçöl, 2012). The notion of nested systems is therefore not just a matter of defining smaller systems within bigger systems, or even assorting them according to 'lower' or 'higher' levels. The idea of 'nested systems' means that different system representations of governmental organization, as developed by individuals and by groups of individuals, intersect in practice and are equally true when actors act accordingly. No wonder, then, that there can be much confusion about who ought to do what with whom about something in the public realm.

What about hierarchies? Legally-defined hierarchies are important constructs in public decision-making. For example, their arrangements imply that a minister can ask his civil servants what to do instead of the other way around. A social systems approach would define a hierarchy as a means of structuring complexity, making use of the consent between actors to act in certain ways. Simon (1962, in Cilliers, 2001) gives three reasons why hierarchies exist in complex systems. Firstly, hierarchies are to some degree efficient. They allow the quick assembly of an (governmental) organization or an organizational field without having to reinvent it all over again. Secondly, hierarchies allow efficient communication because they have established routes for messages. For example, it is essential for governmental organizations that the path in which information flows is clear, so that it arrives at the right destination such as a minister or a committee. Thirdly, hierarchical systems may build up redundancy, which may help to build resilience as elements and functions can replace those that fail. Such hierarchies, in short, contribute to a system's survival (Morçöl and Wachhaus, 2009) but it should be emphasised that hierarchies emerge over time and are not always designed or superimposed.

A hierarchical structure is not necessarily a pyramid with aligned functional relationships. Instead, actors or a group of actors display a mix of intersecting or interpenetrating relationships. A complex systems approach holds that "... the structure of a complex system cannot be described merely in terms of clearly defined hierarchies. This is because the structure of complexity is usually fractal (whole mirroring the parts – LG), there is structure on all scales (i.e. actor-bound definitions – LG)" (Cilliers, 2001: 7). Hierarchies are important for keeping some stability in communication and function, but they intersect in several ways and are tied-in with horizontal relationships. The neat hierarchical division assumed in many models between the realm of control where decisions are made, and the realm of implementation where decisions are carried out is not readily observed in reality. Decision-makers switch between vertical and horizontal relationships and wander in and out of the arena (Cyert and March, 1992; March, 1994).

This view implies that government systems serve as a means to structure complexity, and make it more manageable to actors. Continuous debates and an apparent lack of progress in public decision-making can be partly traced to clashing mental models and differing ideas about what an actor ought to do. Boundaries and hierarchies help structure complexity but at the same time, they can impose restrictions on what can be achieved and how.

2.4.1 A FIRST CONCEPTUALIZATION OF COMPLEX ADAPTIVE SYSTEMS IN PUBLIC ADMINISTRATION

Before returning to the issue of complex adaptive systems, let us briefly revisit the argument so far. First, we have established the need for systemic thinking and theorizing, and reviewed a number of approaches put forth by prominent authors. Early attempts focused on systems 'out there', whereas later attempts pointed out that systems exist in the eye of the beholder. Both efforts have their merits, and we attempt to reconcile both strands to the extent that we accept that system definitions are social constructs but not to the extent that we venture into postmodernist territory where nothing is real. Physical objects may be assigned meaning through communication, but they are real and tangible regardless of the communications surrounding them. For example, while the changes in sediment movement in the Westerschelde estuary are real and measurable, the systemic explanation for such changes is always a social construction. Nonetheless, such system demarcation is necessary in order to be able to arrive at conclusions as it is extremely hard to map the full physical system and all the variables that determine it.

Next, we looked at the systemic nature of public decision-making. We demonstrated the relationship between decision-making and the environment through the work of Easton, and the functioning of policy arenas through the works of Rhodes and others. Governments are structured through boundary judgments and hierarchies, even though the actual hierarchies are not always the same ones indicated on an organization chart. As social constructs, hierarchies become real because people act accordingly (at least until they decide not to do so anymore).

The remainder of the chapter presents a first conceptualization of complex adaptive systems in public decision-making using examples from a study of the operation of the Dutch railway system. This conceptualization is by no means complete, but is a starting point on which the arguments of the rest of this book will be built. We will revisit this conceptualization in Chapter 5. As mentioned before, many meanings and definitions of complex adaptive systems exist. However, those definitions have some commonalities. Prototype definitions, such as the ones by John Holland (1992; 1995; 2006), focus on a surprisingly small number of basic but important dimensions. We identify three principal properties common to most definitions. First, systems emerge through interactions between heterogeneous actors, but not by superimposed control or deliberate design. This means that if actors engage in repeated interaction of any type, structures and processes will emerge and become (somewhat) persistent. Over time, a system will be established out of those interactions (cf. Schelling, 1978a; 1978b; Arthur et al., 1997; Room, 2011a; 2011b). Axelrod, for example, demonstrates how behavioral norms come naturally into existence when people interact in groups, without someone having to superimpose such norms from the outside (1984; 1986; 1997). Thus, social systems do not exist a priori, but are truly emergent, although it should be understood that once a system has come into being, new members will have to conform to established norms.

The second common aspect of complex adaptive system is the adaptive nature of actors in such systems. Structure, processes and norms are not cast in iron, but are the collective responses of individual actors to the incentives presented from within and outside the system. Actors continuously assess their performance in the face of these incentives to determine which behavioral rules promote an improved 'fitness' with the system. Clues about the level of performance, however, are not always clear-cut, so experimentation and adjustment, or adaptation, are integral characteristics of the system's constituent actors. Over time, actors develop mental models of behavioral rules, as pointed out by Checkland (1981) and March (1994), the latter comparing the logic of consequences with the logic of appropriateness. The mental models also include maps of how actors think the system operates because they need to structure and simplify the complexity of their environment (Forrest, 2008). Thus, actors in systems are not passive processors of information but actively seek to gain better positions. If a system emerges out of interactions, it follows that it becomes *dynamic* and *adaptive* through the aggregate behavior of its constituent actors.

The third aspect concerns the non-linear relationship between what happens at the level of the individual actors and the behavior of the entire system. The adaptive moves of actors are mutually different as each actor has a (somewhat) different criterion for what constitutes a better individual fit. These adaptive moves shape the system's behavior but the two do not share a linear relationship. The system is the aggregate of all individual adaptive moves at the individual level, as is its evolution (Flood & Jackson, 1991; Gell-Mann, 1995; 1996; Room, 2011a).

These three basic properties can result in the emergence and development of intricate and dynamic systems over time (Teisman, 2005). Although they provide a starting point for the other arguments in this book, we need to remind ourselves that a CAS is just a first step in understanding complexity. For example, Holland is self-conscious in understanding that his work on CAS represents abstractions that operate as heuristic devices that allow us to start thinking about social reality (2006). In the next section, we will look more closely at what this means for public decision-making by introducing the CAS-based concept of policy action systems.

2.4.2 DRAWING LINES

The current theoretical exercise calls for a good example. To this end, we study the organization and operation of the Dutch railway system (Maasland, Koppenjan & Gerrits, 2011; Gerrits & Koppenjan, 2011). As with other European railways, the Dutch railway system was nationalized after World War II and was run by the state using public funds under the banner of the Nederlandse Spoorwegen (NS). However, increased deregulation, a turn to free-market thinking and spill-over from EU-regulations placed increased pressure on the system. It was thought that the move towards a common European market would also lead to shared railway standards across European nations that would make it easier for trains, passengers and goods to cross borders. While most national companies had their own signaling system, power system, driving rules and so on in the past, it was clear that a unification of such items would help establish a multi-accessible railway network (Vromans, 2005).

Naturally, ease of access and deregulation opened the door for entrepreneurship and private enterprises started to exploit the possibilities of a more or less open network. For example, rolling stock manufacturers offered multi-system locomotives, leasing companies bought fleets of such rolling stock, and (new) train operators leased this rolling stock for their operations. The United Kingdom was well advanced in this respect when a series of major railway accidents, caused by underinvestment in the infrastructure and its maintenance, highlighted the disadvantages of privatization (Pollitt & Smith, 2001).

The developments in the UK halted similar developments in the Netherlands. Train operations (now NS Reizigers) and infrastructure development and maintenance operations (now ProRail) were separated, but neither company was scheduled for full privatization. Still, the minister of public transport saw the need for experimentation, partly because of pressure from the EU, and partly because of a desire to cut costs. Minor feeder lines in rural areas where NSR had lost interest were put out to tender, and companies such as Arriva, Veolia, Syntus and Connexxion won concessions to operate trains (Vromans, 2005). In most cases, those lines experienced a revival with new rolling stock, more frequent trips, increased ridership and higher satisfaction ratings from passengers. Bolstered by this success, the operators united under the banner of the Federatie Mobiliteitsbedrijven Nederland (FMN) and proposed that the minister place even more railway lines out to tender. Their cause was backed by regional authorities who welcomed the revitalization of the railways in their region. This proposal was presented in 2011.

The situation makes an interesting case for system analysis. A railway network, as it name implies, is a tangible

Tickets
VEOLIA
THALYS
THALYS
SPURT
Experts in Railwa Logistics
VEOLIA

system. It has steel tracks connecting one place to the other. The boundaries around this system are determined by decisions made on technical factors, e.g. a different gauge or signaling system. Such boundaries exist between countries and sometimes within countries, as is the case in Switzerland. Although the Netherlands has a technically unified railway system, fragmentation still occurs in many ways. For the sake of control, both NSR and ProRail have divided the network into different and partly diverging sectors that are responsible for specific parts or regions of the network (Jespersen-Groth, Potthoff, Clausen, Huisman, Kroon, Maróti & Nielsen, 2009). There are also smaller operators who operate a number of single lines. These operators may be independent from each other as organizational units, but not independent when it comes to operations. Although different trains share the same network, railways do not allow for multiple trains running the same route at the same time, unlike roads for cars. Operators thus had to coordinate their timetables with each other and with ProRail, in case of delays, failures or accidents.

Arguably, the main aim and reason for the existence of train operators, traffic control, maintenance companies and financing organizations is to offer passengers seamless travel. Passengers do not generally care for technical systems, track maintenance or for a particular company to operate their trains. They want to travel as quickly as possible and as cheaply as possible with minimum fuss. Consequently, there needs to be a unified ticketing system that prevents passengers from having to buy multiple tickets for one trip. Operators need a platform with which to coordinate such a system. Customer demand also requires that NSR, which was previously the monopolistic operator, share its ticket vending machines with other operators and that staff at ticket counters can serve passengers on all lines. Stations should accommodate different operators equally and information systems, including printed time tables and digitals monitors indicating arrivals and departure times should display information from all operators

Another dimension of the system concerns financing and administration. Dutch railways are subsidized by public funds, which means that companies applying for a tender are given an approximate budget to work with. Those who can offer the most value for the budget are granted the concession. The subsidy comes from the national government, but the budget is allocated through regional authorities in the case of lines that are not part of the NSR core-network. Authorities soon discovered that the budget gave them a powerful tool for negotiating deals that benefit the region. But in some cases, the lines put out to tender did not match administrative borders, necessitating further administrative coordination between regional authorities. Another complicating issue is that of reimbursement. Operators pay a fee for track maintenance and signaling to ProRail, but they also receive compensation for each passenger they transport. The compensation has to be paid out fairly, which again necessitates coordination among operators and with the authorities. The fact that NSR used to handle the distribution of funds is a complicating factor.

This is a sketchy first attempt to reconstruct the tangible and intangible interconnected components of the Dutch railway system. A quick glance reveals that the system actually consists of various intersecting subsystems, such as the daily operations system, the planning and control system and the finance system. Closer inspection reveals even more intersecting systems, each with its own boundary. Each anti-reductionist approach to the railway system will point to specific system boundaries, while also including elements of other possible definitions in the quest to make the definition as good a representation as possible. For example, one cannot define the daily operations system without taking into account aspects of planning, which in turn requires the availability of rolling stock.

Systems and system definitions are therefore recursive (Cilliers, 2001), that is, the parts mirror the whole but each definition has its own focus and each can differ greatly from the others. Operators approach the system from the vantage of the lines that they operate, regional authorities approach it from the perspective of administrative division, task specification and accountability, passengers look at the ease and comfort of their trips, and traffic control is concerned mainly with the safety margins of the flows. Each perspective leads to a system definition and a specific logic about what entails a well-functioning system, such that seamless traveling is not easily achieved. The complexity of such a case may thus be better understood when it is defined as a complex adaptive system.

2.4.3 CAS AS POLICY ACTION SYSTEMS

We established in Section 2.4.1 that complex adaptive systems have three main determining properties: structure, processes and norms originating in horizontal interactions; change originating from the adaptive movements of constituent actors; and the system's apparent capriciousness stems from the non-linear aggregation of these adaptive movements. Looking at the Dutch railway system, it appears that decades of NS monopoly have created a system defined by long-standing standard operating procedures and norms. In practice, this implied that it was the NS who decided when and how to run the trains, where to invest in new infrastructure and acquire new trains, and when and how to schedule and execute track maintenance. A change in the political frame of mind toward greater privatization and the promulgation of EU legislation requiring the creation of a common rail market put that existing system under pressure.

The system started to change. For instance, NS sold its cargo department to become NSR, and expanded in Germany, Czech Republic and the United Kingdom through its subsidiary Abellio. At the same time, it aimed to maintain its position on the Dutch network, only discarding minor lines that it saw as unprofitable and arguing that the main lines should never be put out to tender. At the same time, the changing economic and political environment offered newcomers a chance to challenge the NSR with a remarkable degree of success. They assembled in the FMN and made a concerted effort to urge the Minister to put more lines out to tender.

A crude distinction may be drawn between actors who strive for the continuous existence of the old situation, and actors who attempt to change it to a new situation. Both approaches can be considered adaptive in response to external incentives, and to what other actors in the system do. The system as a whole changes in a certain direction as an aggregate of all adaptive moves in different directions. None of the actors can fully align the system with their own adaptive behavior. Consequently, they experience the system as badly managed, poorly understood, and erratic, which points at the non-linear property of the CAS.

Of course, the description above is only a first approximation of how a CAS operates. We also need to take into account its recursive nature and the blurring of boundaries. The CAS of the Dutch railway system is both tangible (e.g. the infrastructure) and intangible (e.g. the coordination among operators). We need to remind ourselves of the discussion in Section 2.2.3 that demonstrated that perceptions and the act of communication and boundary setting are as important in the workings of the system as its more tangible aspects. It is thus necessary to merge the two aspects if we want to understand CAS better. Among others, Crozier (1964), Clark and Crossland (1985) and Hinssen (1998) have pointed at that conjunction and proposed the use of the denominator 'action' to indicate that such systems encompass diverse actors who attempt to survive in the system they are in, as defined through their own boundary judgments and mental maps. This means that they may talk, think and act in the same system, while their operations are not necessarily converging in a common policy arena. The Dutch railway system shows how actors operate in one and the same system whilst at the same time forming and defending different and sometimes intersecting or even conflicting boundaries. For instance, traffic control uses different but intersecting boundaries than train operators do in their daily operations. More conflicting boundaries can be found where NSR and the FNM companies try to gain or defend market share from each other.

We borrow this notion of action systems and extend it to propose the concept of the *policy action system* (Gerrits, 2008; 2010; 2011) in order to explain how actors as members of a CAS converge and diverge over time. A policy action system can be thought of as a subset of the CAS, bearing in mind that it represents a particular section of the CAS but not a different system. The action systems consist of multiple actors and can grow and shrink, merge and subdivide over time (Flood, 1999), depending on what mental maps and system boundaries actors employ. The decisive factor here is the measure of homogeneity or heterogeneity among all actors in the CAS.

A *homogeneous* policy action system generates singularity in the way it handles the incentives of pressures it is subject to. This becomes operational when actors opt to connect and work exclusively with those who support and promote the goals and means already present in the policy action system.

A clear distinction is maintained between those actors who support the current state and those who question it, the latter being shielded away. Singularity in response to complexity is informed by the idea that decision-making is complex enough as it is and anything that stretches beyond the existing system state is seen to complicate matters even further. Such 'distractions' are thought to be potentially disruptive to the current state. Connecting with actors who oppose that state or who offer alternatives to it is seen as gaining an undesirable opportunity to channel unwanted ideas into the action system. Such interaction is seen as a threat to the system, and its members therefore prevent actors with diverging ideas from connecting, out of self-preservation. Singularity can also be observed with regard to the scope of the space of problem definitions and possible solutions. It means that a very limited variety of options are explored and homogeneity results. A homogeneous policy action system draws clear and impermeable boundaries that are only shared among like-minded actors.

A *heterogeneous* policy action system responds to pressures by generating diversity in actors, problem definitions and possible solutions. The system cultivates connections not only with proponents but also with opponents and tries to incorporate the diversity of goals, ideas, and beliefs into the operation of the policy action system. Increased heterogeneity leads to a widening of the scope as actors channel their diverging views and ideas into the system. The resulting fusion of existing and new views brings about a more diverse orientation on the space of problem definitions and solutions. Existing ideas are the subject of continuous debate between adversaries rather than supporters of the same goal. Established ideas are questioned and may be replaced if an alternative, that is perceived to be an improvement over earlier ideas, is offered. Boundaries blur in such systems and complexity is met with complexity as a result. This may improve the substantive quality but, more pragmatically, can also help to garner more support for it from others. Whilst heterogeneous policy action systems are better able to account for the diversity present in the CAS, they are also at risk of generating so much diversity that nothing substantial results from that.

The two variants of policy actions systems presented here are archetypes on one continuum or gradient. They form subsets of the CAS and as such help explaining the clustering of actors across the CAS according to the measure of homogeneity or heterogeneity. Both types of systems have a capacity to reinforce, something we will look at in greater detail when we discuss autopoiesis and self-organization in Chapter 4. It means that a CAS has no natural point of gravity around which all the actors move, no matter how much individual actors may like to think that they are the center whom others should obey. The quest for homogeneity or heterogeneity means that adaptive moves are made continuously, resulting in the non-linear evolution of the CAS as a whole.

2.5 CONCLUSIONS

This chapter set out to explore what is systemic about public decision-making, why systems matter, and how systemic thinking and systems theory deepen our understanding of decision-making. As such, it has a broad scope, revisiting a number of important authors and the theoretical evolution of systems. The chapter also discussed the particular takes on systems within the realm of public administration. While early versions attempted to develop a generic, all-encompassing and functionalist approach to social systems, later iterations focused on the social construction of system boundaries and the process of semiotics that takes place within systems. The concept of the complex adaptive system helps us understand both the systemic and complex nature of reality, because it focuses on the relationship between the adaptive behavior of individual actors and the non-linear relationships they have with the evolution of the system as a whole. This was then juxtaposed with a discussion of semiotics and learning and resulting boundaries, leading to the concept of policy action systems. Such systems are groupings of actors in the CAS according to their degree of homogeneity or heterogeneity in the way they develop their mental maps and system boundaries and act accordingly.

Thinking in terms of systems is the basis for the remainder of this book's argument. Obviously, to research a case such as the Dutch railway system as a CAS requires more precise definitions and a more careful mapping of the patterns of incentives and subsequent responses than was carried out here. However, the conceptualization of the policy action system in this chapter is a first step in the integration of both tangible and intangible complex systems and public decision-making, and we will expand on it in the following chapters. The mapping of the patterns of incentives and subsequent responses implies the examination of feedback loops. Systems become dynamic and complex not just by the properties of their constituent parts, but above all, by the feedback that surges through the system. The next chapter will concentrate therefore on those feedback loops and the ensuing dynamics.

CHAPTER 3.

BEING DYNAMIC: BETWEEN INERTIA AND RANDOMNESS.

3.1

INTRODUCTION

It seems somewhat redundant to have a specific chapter on dynamism in a book that promotes it as the key factor for understanding public decision-making, almost as if the chapter was a fractal expression. However, we want to move beyond the truism that the world is dynamic and interconnected to uncover the mechanisms that make it so. This chapter's main argument is that dynamics occur in both the systems that public decision-makers attempt to govern and in the public decision-making processes themselves. Easton's systemic models of public decision-making, as discussed in the previous chapter, and the models inspired by his work did not pay enough attention to the fact that the input-throughput-output sequence is not a smooth process, but rather an erratic one with multiple continuous relationships between the decision-making context and its environment (Stone, 1987; Sabatier & Jenkins-Smith, 1993). The continuous interaction between what decision-makers want and the actual outcomes could lead to intended and unintended consequences, or even an apparent lack of consequences.

Opening up the black box of causality first mentioned in Chapter 1 to identify the underlying mechanisms will clarify the relationships between decisions and incentives on the one hand, and all sorts of outcomes on the other, both foreseen and unforeseen, desired and undesired. This may help explain, for example, why a project leader may feel that her task is no closer to completion, despite having worked on it for over a year. Alternatively, it may also explain how seemingly minor incentives can lead to major changes.

For example, Hamdi (2004) recounts how the simple act of rerouting a bus line so that it stopped at an intersection in a run-down part of a town led to the establishment of a small informal market. The logic was that passengers might want to eat or drink something while waiting for the bus. An entrepreneur took advantage of the informal market by buying fish at it and transporting it on his bicycle to the middle-class parts of the town where he sold the fish. His modest profits were reinvested in his business, enabling him to expand to a fleet of twelve bicycles, run by his friends. Naturally, they all assembled at the bus stop to load the fish on their trays. Consequently, bicycle repair shops started appearing there. This cycle of economic spin-offs continued, showing how certain effects were self-reinforcing, allowing the creation of increasing returns among a steady group of fish-sellers. It is worth noting that the initial local conditions have a significant impact on outcomes. The bus-stop location was chosen by Hamdi partly because there was a freshwater pump nearby, and this contributed to the success of the fish vendors. Had the pump not been there, the story may have unfolded differently.

Various mechanisms explain the dynamics of the bus stop case study. The aim of this chapter is to identify these mechanisms, demonstrate how they operate, and explore their reciprocal impact on public decision-making. We first introduce a single case study about the planning of port extensions in Hamburg, Germany (Section 3.1.2). This serves as a case *pars pro toto* because it demonstrates the mechanisms we will discuss in this chapter. We then discuss three basic mechanisms of dynamic systems: positive and negative feedback loops (Section 3.2.1) and non-ergodic chance events (Section 3.2.2). These are relative simple mechanisms, whose effects become much more complicated in combination. We therefore need to look at how different combinations of feedback loops lead to non-linear dynamics at the system level (Section 3.3.1). In particular, we need to gain a more in-depth understanding of how systems can be stable or changeable and how they change between the two states. We will therefore first focus on punctuated equilibrium and hysteresis, two related concepts that explain system changes (Section 3.3.2). Next, we discuss path-dependency and lock-in, which explain inertia or stasis in systems (Section 3.3.3). The last aspect that help us understand system dynamics concerns a system's carrying capacity, that is, the extent to which it is able to withstand and recover from the pressures and shocks it is subjected to (Section 3.3.4). We conclude with some final thoughts about dynamics and social systems (Section 3.4).

3.1.2
CASE *PARS PRO TOTO*: THE EXTENSION OF THE PORT OF HAMBURG

It is useful to use a case *pars pro toto* to illustrate the different aspects of dynamic systems presented in this chapter. For this, we to turn to the port of Hamburg in Germany (see Gerrits, 2008; 2011, for a more detailed narrative). This port is one of Europe's largest seaports, and a key asset of north-west Germany in terms of jobs and economic growth. European ports are competing fiercely for market share and extending a port's capacity is seen as the most important strategy for staying ahead of its rivals. Hamburg's port lies approximately 100 kilometers away from the North Sea, making the Unterelbe river and estuary that connects the sea with the port an obvious bottleneck. As it is relatively shallow, the number of ships that can reach the port is restricted. Larger cargo ships have to wait for the tide to rise, providing them with a limited window for arriving at or leaving the port. The largest ships cannot call at the port at all. With the shipping industry building increasingly larger ships, the Hamburg Port Authority (HPA) is constantly seeking to deepen the navigation channel in the river, and has done so since about 1900.

After the 1980 deepening operation, another such operation was planned to for the period between 1995 and 2000. However, unlike the previous occasions, the planning process this time met with considerable resistance from environmental pressure groups and concerned citizens. It had thus stalled by the mid-1990s. There was also opposition from the neighboring federal states Niedersachsen and Schleswig-Holstein. They were worried that they would have had to pay compensation for any environmental damage, instead of Hamburg, because the Unterelbe runs through their territory. Fearing that such resistance would hamper the deepening operation, the HPA and the Hamburg Senate tried to shield the planning process from the public and speeded it up as much as they could. For example, the deepening operation was approved before the obligatory environmental impact assessment was drafted. When this assessment was presented in 1997, it stated that the operation was not likely to cause any unfavorable consequences for the environment. A monitoring program was to be established

BLOHM+VOSS DOCK ELB

to monitor the effects of the deepening. It was expected that the first results would become visible approximately 15 years after the deepening of the channel. While the EIA marked the start of the formal planning process, including the ability to object to the deepening, the HPA had already begun a 'preparatory dredging' operation. This inevitably angered its opponents, most notably Niedersachsen and Schleswig-Holstein. Nevertheless, the operation went ahead and was concluded on December 14th, 1998. Ships with a draught of up to 12.80 metres were now able to call at the port without being dependent on the tide.

The prospect of a booming market for cargo shipping made it tempting to pursue a further deepening of the Unterelbe. However, Hamburg had calmed the opposition by promising not to plan a new deepening operation any time soon. This promise was crossed by a political change that brought the Christian Democratic Party (CDU) to power in traditionally social-democratic Hamburg. As an opposition party, the CDU had always opposed the then-current Senate's promise not the deepen the Unterelbe again. Now that the CDU had grabbed the power, the planning process was sped up, and a new deepening operation was announced in April 2002. At this stage, the environmental consequences of the previous deepening operation were unknown, as it was still too early for the consequences to appear in the data. Some stakeholders argued that the HPA should delay the planning of a new dredging operation until the first monitoring report, which was not due till 2013, became available. However, others argued that the lack of observable unfavorable consequences was evidence that nothing wrong had happened and that the next channel deepening operation could go ahead.

However, while the public decision-making process for the channel deepening operation was was being finalized, the dredgers suddenly faced a major shift in the Unterelbe. They found that the volume of material dredged during the maintenance operations in 2004 was considerably more than that dredged previously (see figure 3.1). Soundings confirmed their observation that the volume of sediment flowing in from the North Sea and accumulating in the harbor basin had suddenly increased from the usual 4.5 million cubic metres in 2003 to 9 million cubic metres thereafter. This was an unpleasant surprise, as the port authorities now faced escalating costs because of the need to dredge the port continuously as to avoid the port from silting up. Another problem was the lack of space for disposing the dredged material. Without any space in the city, Hamburg turned to its neighboring states. However, they were unwilling to cooperate because it had offended them by pushing through the previous deepening operation. HPA therefore chose the only option left: to take the sediment to the border of its territory, dump it in the Unterelbe, and hope that the tidal currents would then take the sediment to the North Sea. However, this does not take place, because the sediment originates from the North Sea in the first place. Thus, the HPA has to continuously dredge the port in an almost vicious dredging cycle.

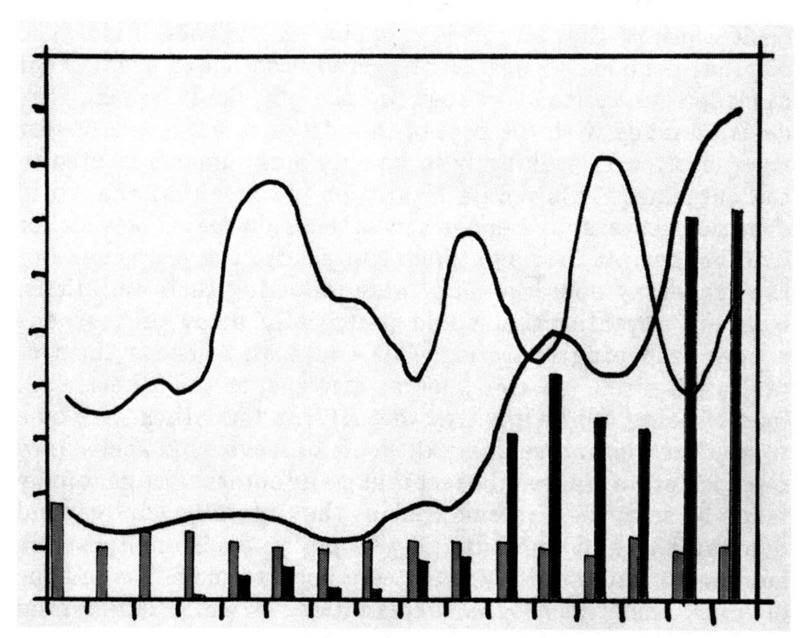

Figure 3.1: Sediment accumulation in the Unterelbe between 1990 – 2005. The total volume of sediments is indicated by the black line. The total volume comprises sediments that are processed locally, e.g. storage, remediation (grey columns), and sediments that are disposed at the territorial borders of the City of Hamburg (black columns). The grey line indicates the fresh water discharge at Neu Darchau (Gerrits, 2008, adapted from Bundesanstalt fur Wasserbau, 2005).

The sudden and sustained increase in sedimentation in the port can be attributed to the deepening operation. By lowering the riverbed, the tidal regime shifted from a system where the fresh water discharge was larger than the influx of water from the North Sea, meaning that the sediment would be flushed out to the sea, to a flood-dominant regime, where increased room in the riverbed allowed sea water to transport the sediments the other way around and back to the port. One may argue that this change could be reversed by putting back the last layer of sediments that were removed. However, that turned out to be very difficult because the flood dominance of the new situation obstructed such an operation. Consequently, engineers devised new solutions such as building shoals in the river's mouth to slow down the tidal currents, and constructing a sub-aquatic sediment disposal site.

To summarize: the results of the deepening operation went against the very raison d'être of the operation. The original decision to deepen triggered a series of events that cascaded to cause a change in the physical system. The decision made with the best of intentions to achieve a deeper river was found eventually to be only ambiguously related to the outcomes. This points clearly to the fact that the world does not necessarily behave as the decision-maker wishes for it to behave. At first sight, one can accuse the proponents of the deepening operation of blindly following their ambitions, ignoring anything that could potentially delay or postpone a new deepening operation. One could also accuse them of faulty research, as the popular newspaper headlines were fond of doing. While it is true that HPA and its allies were on a route that was increasingly difficult to leave, this accusatory interpretation ignored the fact that many actors were genuinely taken by surprise. Systems such as the Unterelbe can respond dynamically and unpredictably to policy decisions, possibly leading to unfavorable results. We need to move beyond the obvious, sometimes banal explanations to fully understand the mechanisms behind these surprising results.

3.2.1
FEEDBACK

The Hamburg case is an example of multiple interacting systems. The first was a physical system, the Unterelbe, that changed to a flood-dominant river because of the dredging operations that were carried out. In response, these physical changes required decision-makers, operating in a rather homogenous policy action system, to respond with emergency measures, such as cyclic dredging and the construction of artificial shoals in the estuary's mouth to slow down the tidal currents, which amounted to yet another physical change. The pressures on the policy action system, following the sudden increase in the volume of sedimentation, also caused changes in the policy action system. Some actors began questioning the way the operation was pushed through and wondered whether they had been open to alternative options. Various ideas, including those offered by environmental pressure groups, were used to explore other ways of dealing with the Unterelbe, and the policy action system grew more heterogeneous as a result.

In short, what we are dealing with in this case is the continuous interaction - in the shape of actions and subsequent responses - between a physical system and a policy action system that leads to far-reaching changes in both systems. Dechert describes such patterns in public decision-making as mutual causal processes, a term that can be applied to describe "those activities or states of a system or of its components which are contingent upon interaction with other systems (environmental system) or other components of the same system and in which "the size of influence in one direction has an effect on the size of the influence in the other direction and is in turn affected by it. . . . " Such mutual causal processes may be deviation-counteracting, resulting in an equilibrium based on negative feedback; or they may be deviation-amplifying processes based on positive feedback in which there is increasing divergence from the initial condition." (1966: 10). Thus, if we talk about reciprocal or mutual causal processes – where one element impacts another element and vice versa – we talk about feedback loops.

Positive and negative feedback are the source of *all* dynamics in complex systems (Sterman, 2000). A positive feedback loop occurs when the response to a given input or incentive is reinforcing or amplifying, and in some cases is disproportional to the input (Marion, 1999). Positive feedback loops therefore lead to change in systems. Note that a positive feedback loop could lead to both favorable and unfavorable changes. That is, the word 'positive' does not indicate that something positive has happened; the direction does not matter for the qualification. Recall the example of urban degeneration from Chapter 1: the reinforcing factors of suburbanization and less attractive city centers were in a positive feedback loop, with both effects reinforcing each other, leading to unfavorable results such as abandoned urban cores.

A positive feedback loop also occurred in the Hamburg case. The previous deepening operation reinforced the natural processes of sediment transport and sedimentation in the Unterelbe, to such an extent that the system shifted to a new stable, but unfavorable, state as a flood-dominant sediment transport system. Positive feedback loops can also bring about favorable results, such as in the case of competing technologies where one technology becomes an increasingly dominant force in the market, thus benefiting its patent-holder and manufacturer (Arthur, 1994). Urban degeneration may become urban regeneration if, for instance, squatters begin to run a breakfast and coffee bar in a derelict building, attracting people to that area and perhaps persuading them to start adjacent businesses.

A negative feedback loop is a self-correcting loop that has a dampening and stabilizing effect (Marion, 1999). The response to a certain input or incentive is to return to the previous state, so that it corrects the disturbance that occurred by closing the gap between the old and new situation (Diehl & Sterman, 1995). One could argue that the deepening operations that were carried out in Hamburg prior to the operation of 1998 were part of a negative feedback loop. Although the riverbed was lowered because of the dredging works, the sediment transport and sedimentation restored itself to the previous situation. This meant that the sediment volume in the Unterelbe did not increase significantly for a long time. Positive and negative feedback loops can occur simultaneously, sequentially and on different timescales (Diehl & Sterman, 1995). They are not each other's opposites, and are instead destabilizing and stabilizing forms of causal relationships. Figure 3.2 depicts these two forms of feedback.

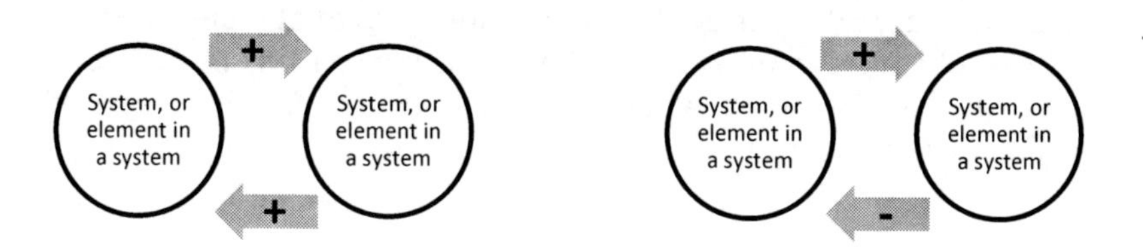

Figure 3.2: A positive feedback loop (left) is self-reinforcing; a negative feedback loop (right) is self-correcting.

Feedback between systems and between elements in systems causes incentives and adaptation, leading to changing or stable situations. For example, Norgaard (1994; 1995; Gual & Norgaard, 2010) showed how the optimization of physical systems for human use, e.g. in agriculture, has in fact created an intricate pattern of feedback loops that not only changed physical systems but also pushed public decision-makers to create increasingly individualized task specifications and more complex institutional and cultural contingencies to deal with the ensuing feedback. Over time, the complexity of this pattern renders it nearly impossible to attribute any particular development to a specific feedback loop, as the two types of systems have become completely intertwined. It would therefore be overly simplified to say that there is only feedback loop between systems. In everyday situations, it is likely that many of the loops take place simultaneously. Feedback may result because of intended action but could also result from unintended actions, or lack of actions.

The Hamburg case is a useful example because the reciprocal inputs and responses are tangible. They appeared in the data points in various reports and one could even see the sediments in the city's canal during the ebb periods. This was a case of *material* feedback loops. Since we are also dealing with intangible social systems in general and policy action systems in particular, we should note that many of the feedback loops in the world of public decision-making concern *information*, which is often more ambiguous to trace and map. Feedback in the shape of information flows means that people receive, interpret and process information using mental models, consequently leading to (in-)actions in response to the received information. While their responses might be reinforcing or dampening, these effects may be harder to establish because of their intangibility. It is also worth noting that the perspective of the observer matters in qualifying something as positive or negative feedback when feedback occurs in the form of information.

3.2.2 NON-ERGODIC CHANCE EVENTS

Different feedback loops can render surprising outcomes. The change in the Unterelbe provides an example of such a surprise. The port authorities had been carrying out deepening operations since 1900, with roughly the same result each time. However, the deepening operation in 1998 suddenly led to a novel outcome with real consequences, an indication of the rate of the feedback loops. While previous operations did not affect the main processes of sedimentation and sediment transport, the last one did, causing a set of factors that had not previously influenced each other to do so. It came as a surprise to the decision-makers who may have had counted on a repeat of prior outcomes. If someone had been able to foresee the new combination of factors, the new event could also have been predicted. However, that proved something impossible. It was not lack of foresight, because much research had been carried out, but rather an impossibility of foresight that was the cause of the problem. Novel combinations of factors are important for understanding the dynamics of complex systems. In fact, social complexity is partly "about how a seemingly trivial event may trigger changes far removed from it in time and space." (Giddens, 1984: 10). At this point, our argument splits in two directions: first, the occurrence of chance in complex systems; and second the subjective experience of chance and resulting uncertainty at the level of the individual decision-maker. We will deal with the first issue in this chapter, and the latter in Chapter 4.

Broadly speaking, the purely analytical concept of chance

events is used to mark novel events that happen without an apparent cause and that deviate from what is regular. Emphasis should be put on the word 'apparent' because while each chance event has a cause, people may fail to foresee and observe its cause, which means that it appears to be unexpected. Also, the word 'cause' requires some consideration. It does not imply a universal cause, because something that is universal repeats through time, thus making it regular rather than novel. Instead, chance means that the *conjunction*, the co-occurrence of factors that had not previously come together introduces an element of randomness to the world. As argued in Chapter 6, events and processes are the result of particular configurations of conditions. It is therefore not possible to assign a discrete impact to each individual variable that influences the event or process. A chance event is just such a conjunction of factors, with a novel outcome. That novelty means that this particular conjunction has not occurred before and may not be repeated. That is, its novelty diminishes if it occurs continuously.

Nelson & Winter (1982) propose a categorization of three types of such events to understand the subtle differences between pure chance and apparent chance. *Random* events are both unpredictable and untraceable, so that it is impossible to trace their causes once they have taken place. *Unpredictable* events differ from random events in that it is possible to trace their causes, but only after they have occurred. This means that, even though they can explained post-hoc, they are still unpredictable. *Deliberate* events are novel events that are created with the intention of causing something to change. This distinction draws attention to the fact that an observer's position and response to the chance event matters for its nature and its impact. A particular conjunction may have occurred, but if it was not observed or did not lead to any changes, it is hard to say anything about it. For a chance event to matter, it has to be observed and responded to.

Rescher (1995), among others, draws attention to the fact that what seems to be a chance event from one perspective could be a planned outcome to someone else. Suppose that someone has bought a new house in a neighborhood and moves in on Sunday, only to discover a letter from the public utilities announcing that they will close down the water supply and sewage system in the street from Monday onwards for maintenance. To this person, the occurrence of the maintenance works is unexpected and probably also inconvenient, if not unpleasant. To the utility company, the maintenance is something that was planned ahead and part of their routine operations. The chance event could be just a coincidence if it had generated no consequences. In this example, it can be imagined that the person who is moving could decide to return to her old house for another week, if possible, rent a hotel room, or put up with the inconvenience for a week. In other words, a conjunction of factors - the planned move, the planned maintenance works- had together brought about an unexpected outcome with a lasting impact. Thus, not every event is a chance event.

The issue that matters for our discussion is non-ergodicity. Non-ergodicity refers to the fact that chance events that occur at a certain point in time can have a lasting impact on subsequent developments: "Important influences upon the eventual outcome can be exerted by temporally remote events, including happenings dominated by chance elements rather than systematic forces. Stochastic processes like that do not converge automatically to a fixed-point distribution of outcomes, and are called non-ergodic." (David, 1985: 332) For instance, the unexpected electoral victory of the CDU in Hamburg, driven by a quick change of popular sentiment, meant that a new deepening operation was suddenly back on Hamburg's policy agenda, leading to the planning of a new operation despite the existing problems with sediment transport. This sudden change in political office can be regarded as a chance event. Non-ergodicity explains how the consequences of a chance event resonate across time. It indicates that long-term outcomes are not just the result of continuous developments, but also depend on chance events that occur (Arthur, 1994). Therefore, non-ergodic chance events matter the most if we want to understand how chance impacts system dynamics.

3.3.1 NON-LINEAR DYNAMICS AT THE SYSTEMS LEVEL

So far, we have discussed individual mechanisms in system

dynamics: positive and negative feedback loops, and non-ergodic chance events. It would be very convenient if the dynamics of the system in which decision-makers operate were determined by only one feedback loop because it would enable us to understand the system instantly once we understood that particular feedback loop. A few hypothetical exemptions notwithstanding, this is not the case in the real world. System dynamics are determined by the cumulative effects of the entire set of positive and negative feedback loops operating in a system. What we are interested in is to understand how the feedback loops in a system relate to the dynamics of the system as a whole, since there is no direct relationship between a particular feedback loop and the dynamism of the system. This lack of proportionality between input changes and outcomes is often called non-linearity, a term we already used rather loosely in the previous chapter. While non-linearity is, strictly speaking, a mathematical equation that cannot be solved when it is broken down into its discrete parts (Kiel & Elliott, 2005), we use the term here to refer to the lack of a direct or proportional relationship between individual inputs and the aggregation of those inputs in the overall dynamics of a system.

The obscuring of the relationship between decisions and outcomes is important here because each policy decision essentially represents an intention to use incentives to change something somewhere in a system. Therefore, each decision assumes, either implicitly or explicitly, a specific causality: if we do this or that, the system will respond in this or that way. Inevitably, this so-called policy theory (Hoogerwerf, 1984; Leeuw, 2003) cannot address all of the feedback loops present in a system, which means that there is often no direct relationship between a decision and an outcome. In fact, a decision may accidentally trigger other feedback loops, producing unintended and sometimes even unwanted results. Some examples of this include: attempts to restrict infrastructure budgets leading to cost overruns (Teisman, Westerveld & Hertogh, 2009; Hertogh & Westerveld, 2010), desegregation policies leading to an increase in housing vacancies (Frank & Ganapati, 2008), and, of course, river dredging changing the tidal regime as shown in Section 3.1.2.

Such examples, which are easily found in social science research, show that a certain policy decision and the consequent public policy does not necessarily lead to the desired results because other feedback loops in the system may interfere: "Our attempts to stabilize the system may destabilize it. Our decisions may provoke reactions by others seeking to restore the balance we upset." (Sterman, 2000: 5). Forrester (1971) calls such phenomena "the counterintuitive behavior of systems". Note that the responses to incentives may amount to non-ergodic chance events because attempts to influence the system could create novel conjunctions and lead to new pathways. In addition, such examples focus our attention on the impact of system openness and the consequent contingencies that were discussed in the previous chapter. In short, the latter indicate that what works at a certain time and place may not work in another time and place. For instance, Pressman & Wildavsky's implementation study (1973) demonstrates how a policy developed for a particular system or context led to unintended effects and policy failures when implemented in a different context. The renowned urban planner mentioned in Chapter 6 is another such example. Although her policy ideas and implementation strategies were highly successful in Rotterdam, the very same approach made her persona non grata in another town.

In this book, non-linearity means that there is no direct relationship between a decision and its outcomes, because decisions may face 'policy resistance' (Meadows, 1982) - they may be delayed, amplified, diluted or even defeated. This understanding helps us appreciate the far-reaching consequences of Flyvbjerg's assessment in Chapter 1 that public decision-making takes place in a highly erratic world where the results are often surprisingly different from what was expected. At the same time, this awareness encourages us to abandon the synoptic model of public decision-making in exchange for a systemic one that leaves more room to anticipate the feedback effects described here.

The archetypical patterns of system dynamics that result from different and interacting feedback loops include the s-curve of growth or decline, oscillation, growth with overshoot and overshoot with collapse (see Sterman, 2000 for a concise overview). The accumulation of positive and negative feedback loops can increase the pressure on the current state of the system to either change to a new state or to remain in that particular state. The Hamburg case is an example of how repeated changes to the Unterelbe accumulated over time, without causing significant changes to it initially. However, once a certain threshold was reached, the river suddenly shifted to a new stable state. This example demonstrates that systems do not always change gradually, making it important to understand how irregular or erratic change results from accumulating feedback loops.

3.3.2 PUNCTUATED EQUILIBRIUM AND HYSTERESIS

Two mechanisms that help explain such erratic changes are punctuated equilibrium and hysteresis. Punctuated equilibrium refers to the situation described above: a system remains stable under (increasing) pressure until a certain threshold or tipping point is reached, after which it moves rapidly into a new stable state. The origins of punctuated equilibrium lie in evolutionary biology where it provided a counter-argument to the thesis of gradual and uninterrupted evolution. Among others, Eldredge and Gould (1972) argued that species do not evolve gradually over time but rather during short periods that punctuate long periods of stasis during which no change occurs. Environmental selection is assumed to take place only during these short periods of change. Punctuated equilibrium has been applied in the study of other fields, such as the philosophy of science and industrial innovation (Gersick, 1991). These applications have focused on identifying the periods of stasis that alternate with the short bursts of change in these domains, and the factors that enable and limit the extent of change, which drive the deconstruction and reassembly of a system and its elements during the periods of change (ibid.).

A well-known application of this explanatory mechanism in the realm of public decision-making is Baumgartner and Jones' (1993) study of the role of agenda-setting in shaping and changing policies and associated institutions. They applied punctuated equilibrium to show that policies and institutions do not live forever, but instead change with the policy agenda. Stability occurs when there is consensus about combinations of problems and solutions and their relative priorities in the public agenda. Conversely, instability is generated when new policy issues are introduced, and is caused also by the portrayal of these issues and where specifically these portrayals can be legitimately displayed (Parsons, 1995). The instability phase in the policy action system allows access to the policy agenda and, consequently, the possibility to change it, as well as the institutions that support it. A new period of stability sets in when new issues, the agenda and the supporting institutions are matched with each other.

Baumgartner & Jones' study is one particular application of punctuated equilibrium in the domain of public decision-making. Other examples include policy paradigms and learning (e.g. Hall, 1993), organizational change (Tushman & Anderson, 1986), and policy choice and institutional change (e.g. Room, 2011). The argument in Chapter 2 that systems are open indicates that the evolution of the environment impacts policy action systems and vice versa. Broadly speaking, there are two ways of understanding this. While some researchers focus on the capacity of external chance events to recalibrate systems with their environment, others argue that such recalibration arises out of the growing gap between stable systems and the evolving environment. Once the gap is too large, the system adapts (e.g. Hrebiniak and Joyce, 1985). Sementelli (2007) uses the metaphor of the Red Queen from Alice in Wonderland to demonstrate how systems change to maintain their relative position and not lose ground when their environment evolves. Chance events and widening differences between systems and their environments leads to a build-up of pressure on systems, and after some time, they respond to the pressure by changing. The apparent stability of systems means that public decision-makers may not perceive that they are under stress and that a sudden change is imminent. The location of the threshold, the point at which a system enters a state of turbulence, remains unknown to them until it has passed (Gunderson, 2001; Walker & Meyers, 2004).

Punctuated equilibrium gives us an initial understanding of the transitions between system states. It raises a number of questions: to what extent are new states discontinuous from previous states? How much history is carried over from previous states? Can new states be reversed? These questions point out the need to need to further dissect the transition between system states. Generally speaking, a new system state is a discontinuity in time- the system as such has not changed but has instead moved into a new equilibrium where some of its characteristics or operations are markedly different from the previous equilibrium. Consider, again, the example of the Unterelbe. While there have been profound changes in the system, i.e. the shift to flood-dominant sediment transport, the system as a whole has remained: the river and its channels and shoals are still there. The transition of a system to a new state does not mean that the old system has been completely broken down and rebuilt in a completely new manner. Therefore, in what sense is this a discontinuity?

A system that has undergone a transition carries the traces of its past state into the new state (Elster, 1971; Isaac, 1994). A punctuated change is therefore not a disruption. In theory, this could imply that a system could move back and forth between two or more system states. After all, if the new state embodies traces of its history, the system should be able to retrace its previous state. However, according to the hysteresis effect, restoring a state to its previous state once it has moved into a new state requires much more energy than was required to push it over the threshold into its new stable state (Strogatz, 1994; Scheffer, Carpenter, Foley, Folke & Walker, 2001; Scheffer & Carpenter, 2003; Scheffer & Westley, 2007; Walker & Meyers, 2004). In other words, hysteresis explains "that a transition from one state to the other does not imply that the reverse can occur." (Graham & Seltzer, 1979: 63).

Suppose that a system at time t_0 is put under continuous pressure p. This pressure could be in the form of the regular removal of sediment from the riverbed, popular protest, lobbying or research results, such as those reported by Baumgartner & Jones (1993). Each system has some capacity to absorb these pressures without needing to change its structures and processes, something we will discuss in more detail in Section 3.3.4. For example, multiple sediment layers were removed from the Unterelbe without leading to major changes in the river. Similarly, some regimes have great resilience against popular revolt and maintain their state easily. This pressure can increase for a considerable period of time t_{1-n} where a constant amount of pressure ($1p$) is added at each point in time ($_{1-n}$). A sudden transition of the system state occurs at t_{n+1} at which point it shifts to a new equilibrium. Although it appears that increasing the level of pressure by a single unit (1*p*) caused the system to move to the critical threshold, this is not the case. Instead, the shift was the result of the accumulated pressure during the period before the threshold was reached. A concrete example is that a regime's collapse should be attributed not to the latest popular demonstration, but the cumulative effect of the demonstrations that occurred in the period preceding the regime change. Similarly, the planning process was cut off because of the accumulation of all the set-backs that plagued the process, not just the latest set-back.

While one may argue that system transitions could be undone by reversing the last 1*p*, the hysteresis effect implies that the point at which a system loses its resilience and becomes vulnerable to the pressures that could push it into a new state lies much before the actual tipping point. This point is called the bifurcation point. A reversal to the previous system state occurs only if conditions are reversed far enough to reach the bifurcation point, not the actual threshold or tipping point (Scheffer & Westley, 2007). For instance, the hysteresis effect in the Unterelbe meant that it was very unlikely that replacing the most recently-removed layer of sediment back on the riverbed would reverse the flow of sediment. Instead, engineers developed alternative solutions, such as the construction of artificial shoals to dampen the flood waves and a further lowering of a particular section of the riverbed that would act as a sub-aquatic sediment trap. More recent examples of the hysteresis effect can be found in Scheffer & Westley (2007) and Scheffer & Van Nes (2007).

Hysteresis shows that the past and the present states of systems are related. However, while they are not completely discontinuous, it is also not possible to directly switch back and forth between system states. Since social systems are open, their environments evolve as they change. This is essential because hysteresis and the openness of systems indicate that time is irreversible. Even if an attempt is made to restore the previous state of a system, the result will *not* be the previous state, but a new state that *resembles* the previous state. It is therefore impossible in social systems to return to what once was, an issue discussed in more detail in Chapter 4. Punctuated equilibrium and hysteresis explain the erratic nature of change as driven by feedback loops. However, systems do not change all the time, as they can also display inertia under pressure from the same feedback loops. This can be explained by the concepts of path-dependency and lock-in.

3.3.3 PATH DEPENDENCY AND LOCK-IN

Although they can change, complex systems can also display long periods of stasis, even in the face of efforts (e.g. decisions, policies) to change them. This stability can be explained by the concepts of path dependency and lock-in (Bergh & Gowdy, 2000). Path dependency refers to the impact of the previous state of a system on any changes that occur; in other words, history matters (Greener, 2002; Pierson, 2000), while lock-in represents the inflexibility that exists in a particular context because the amount of energy required to change it exceeds the benefits of preserving it (Arthur, 1994; David, 1985). In the domain of public decision-making, these two concepts imply that it can become increasingly hard to leave a certain route once it has been selected, either intentionally or unintentionally, even though it appears to be unfavourable in the long run compared to an (hypothetical) alternative (Pierson, 2000).

Path-dependency and lock-in are the result of accumulating feedback loops. Paradoxically, these are positive feedback loops, which were said to cause change earlier in the chapter. However, the accumulation of positive feedback loops can lead to stasis in complex systems. The concept of increasing returns is relevant for understanding this. Increasing returns originates in economics where it is used to indicate a situation "when the output of an economic system increases more than proportionally with a rise of input. In other words, the effects of extra input will be magnified, meaning that *positive feedback* prevails." (Den Hartigh, 2005: 2; italics original). Increasing returns originates in economics where for long it "has had a long but uneasy presence in economic analysis" (Arrow in Arthur, 1994: ix) because much economic reasoning is built on the assumption of homeostasis where economic actions constitute negative feedback loops (Arthur, 1994). However, it has re-emerged when some economists began to conceptualize economic systems as complex adaptive systems (cf. David, 1985; Arthur. 1994, 1997).

An example from Gerrits and Marks (2008) will clarify how path dependency and lock-in are relevant for public decision-making. Suppose that public decision-makers have to choose between two policy options (*A* and *B*) that are available at the same time and could potentially solve the same issue, but are also mutually exclusive, i.e. choosing one option rules out the other. Each option in this example displays increasing returns: as more people follow it, the pay-off from it increases. The choice as to which option will be deployed is determined by two factors: the level of expected returns and the occurrence of chance events. In terms of expected returns, the bounded rationality of public decision-makers makes it more likely they will choose the option that they believe will deliver a higher return on investment in the short run, because they are unable to predict the returns of each option in the long run (David, 1985). Good returns might take a monetary form, but could also include electoral support, better healthcare or a higher number of students enrolled at universities. The second factor that influences the choice of a policy option is the occurrence of chance events at the moment of when the choice is being made. These could include the availability of an individual with more knowledge about option *A* than option *B*, a critical article in the local newspaper about option *A*, resulting in public pressure against it, or the occurrence of an accident which has to be responded to. The occurrence of chance events affects which option is chosen, which means that the chance event will resonate through time (the discussion on non-ergodicity in Section 3.2.2 is relevant here). Thus, the actual outcome of the decision-making process is determined by the expected returns and non-ergodic chance events, introducing an element of unpredictability into the process.

By adopting one of the two options in our hypothetical example, the inducement for decision-makers to change to a different option becomes ineffective. Suppose that the search process results in option *A* being preferred over option *B* . This choice makes it attractive for others to follow the choice, because they may perceive it as allowing them to reach their own goals more easily. This enhances the returns to adopting option *A*, leading to a bandwagon effect with other decision-makers following suit. Each time option *A* is chosen, the less likely it becomes that option *B* will be selected. In the long run, this introduces inflexibility and a fixed path based on option *A*. Figure 3.3 maps our example and shows the number of adoption decisions and the returns for each option.

ADOPTIONS	0	10	20	30	40	50	60	70	80	90	100
OPTION A	10	11	12	13	14	15	16	17	18	19	20
OPTION B	7	9	11	13	15	17	19	21	23	25	27

Figure 3.3: Increasing-return adoption payoffs (Gerrits & Marks, 2007). The columns show the expected return for each option through time.

After choosing option *A* thirty times, choosing option *B* next would have led to higher returns. However, option *A* has been adopted by so many individuals and for such a long time that the increasing returns resulting from this choice benefit those who stay on this path, even though they could see how option *B* could have benefited them at a certain point in time. In other words, the costs of leaving option *A* for option *B* are higher than the returns, even though option *B* would have returned better results by now. There is no way to close the gap between the returns of option *A* and the returns of option *B*, which means that the system has become locked into what can be considered an inferior choice (Arthur, 1994).

At this point, it is worth summarizing Gerrits and Marks' (2008) study on the construction of dykes in the Dutch Zeeland delta because it demonstrates how increasing returns and path dependency manifest themselves in public decision-making. Existing water safety policy programs in the Netherlands are concerned about the delta's geometry, specifically the way the dykes in this region are constructed and aligned. This geometry is deemed inadequate because it promotes undesired currents and sediment depletion in areas where it should not and leads to the disappearance of brackish water that is deemed ecologically valuable because of the closure of certain tributaries. The suggested changes to the delta and the associated massive costs raise the question as to why the delta's dykes were ever constructed in the way they were. Were decision-makers short-sighted back then?

Analysis showed that the decisions for the dykes and

land reclamation projects that have been made since 1000 AD till now were locally (i.e. at that moment in time and place) rational and defendable. The choices that were made yielded higher results in the short run. For instance, land reclamation allowed farmers to extend their farmland so that they could grow more crops, which was more profitable to them than unused brackish meadows. In addition, the availability of farmland and stable regular harvests meant that communities in Zeeland began to grow, which necessitated even more land reclamation (Cox, 2003). Here, we witness a positive feedback loop at work: the greater amount of land available increased harvests, which encouraged population growth, and led in turn to higher demand for crops and thus higher rewards for more land reclamation. Chance events mattered much in this history. For example, multiple floods have breached many of the dykes. Those dykes could not always be restored to their original states with the technology and tools available. Consequently, new dykes perpendicular to the old dykes were laid out, instead of bridging the gaps. This situation provided the starting point for further land reclamation. It resulted in an angular overall geometry in the whole delta. Thus, the decisions that were made locally in time and place were locally optimized, reasonable and defendable, if somewhat myopic. But in the very long run, such local choices lead to an overall geometry that is currently deemed unfavourable and can only be rectified at considerable expense.

The example shows that history matters for understanding the dynamics of complex systems. Complex systems follow certain trajectories through time. Positive feedback, such as increasing returns, may trigger many possible future states. However, once a certain option is chosen, there is no way back because of the same feedback loops, even though in the long run other alternatives may be advantageous. Myopic decision-making and non-ergodic chance events determine the pathways chosen, so that seemingly minor factors or issues resonate over time. Complex systems are therefore "sensitive to initial conditions [...] once the system 'chooses' one branch over another and travels sufficiently far along that path, it stabilizes and the system settles into its new evolutionary pathway." (Reed and Harvey, 1992, pp. 363-364)

Path-dependency and lock-in are important for understanding that even very small unpredictable events may cause a system, by optimising at the local level, to get on a path that is practically impossible to leave. There is no real way to overcome this, as both myopia and chance events are inevitable in public decision-making. In this way, positive feedback loops can lead to stable systems. Note that we are looking at an accumulation of feedback loops, meaning that positive feedback loops *could* lead to stable situations, but not necessarily so. Stable situations can also be explained by negative feedback loops (Bergh & Gowdy, 2000). Thus, each analysis needs to define the particular configuration of feedback loops it is examining. After discussing the main mechanisms underlying the dynamics of complex systems, we next explore the capacity of systems to deal with those dynamics.

3.3.4 CARRYING CAPACITY, DISTURBANCES AND RESILIENCE

So far, we have argued that system dynamics are the cumulative result of feedback loops in a system. Not only can feedback loops reinforce or dampen each other, leading to specific systemic dynamics such as inertia through lock-in, they can also cause different effects when operating sequentially over time. Consider, for instance, exponential growth, which is initially fuelled by positive feedback. However, very few things can increase or decrease exponentially ad infinitum. It is very likely that growth will decrease at some point in time, or perhaps even reverse itself, as a system reaches its natural limits. This is an example of the initial positive feedback loop being opposed by a negative feedback loop. A useful example is population growth in cities (cf. Forrester, 1969). Initially, populations grow exponentially because a larger population has a greater birth rate. However, this cannot go on forever in a city. At some point in time, growth slows down because the city's facilities and supplies cannot keep up with the growth. For instance, there might be insufficient clean drinking water, housing or doctors. In such cases, growth is limited by the carrying capacity of the system, i.e. the capacity of the system to sustain further growth. If the system's stock, as discussed in Chapter 2, is replenished or kept up to date, the system can continue to operate as before. If the stock is being used at a faster rate than it is replenished, its carrying capacity is weakened and the growth that relied on it slows down or may even collapse.

It is difficult to identify the point at which developments driven by positive feedback loops are countered by negative feedback loops, or other combinations of loops. There are three main constraints to doing so. First, a great number of feedback loops operate simultaneously in urban systems and they are not neatly separated (Oates, Howrey & Baumol, 1971). Second, there are delays in feedback loops as their effects on a system's carrying capacity may take a while to reveal themselves, and hysteresis may occur. Third, a system's stock is not necessarily stable and may be altered or replenished. In addition, the time-scale over which a system is observed affects the interpretation of the process. While certain developments may appear to be stable or locked-in if observed over some period of time, they may simply constitute temporal stable states between periods of rapid change. This may be captured if the observation period is extended at both ends of the series. For instance, the Unterelbe case study showed how continuous dredging exhausted the system's carrying capacity in the shape of downstream sediment transport, leading to a shift in the system's stable state. We marked that as a major change. Geographically, however, the Unterelbe is just a rather short section of a very large river totalling over 1000 kilometres in length, whose origins date back several millennia (Maring & Gerrits, 2006). From that point of view, the latest change may be considered a little hiccup in the river's history.

Despite these analytical constraints, we need to understand the carrying capacity of systems to appreciate how certain decisions can promote, deplete or restore the carrying capacity of systems. The term 'resilience' is often used to describe the extent to which a system is able of restoring itself when it is subject to pressure. The idea of resilience originates in the environmental sciences (e.g. Holling, 1973) and has become an important factor in research on the workings of ecosystems and the reciprocal relationships of ecosystems with social systems (e.g. Folke, 2006; Folke, Hahn, Olsson & Norberg, 2005; Klein, Smit, Goosen & Hulsbergen, 1998; Kotchen & Young, 2007;

Rammel & Bergh, 2003; Scheffer, Carpenter, Foley, Folke & Walker, 2001). For instance, Jared Diamond argues in 'Collapse' (2005) that Easter Island's inhabitants decimated themselves because they used the island's trees faster than they could grow back. This overuse of the trees also caused soil erosion that exacerbated the deforestation. Although his explanation has been contested, it highlights the real impact that human decision-making can have in the feedback processes that shape a system. Other examples include the collapse of financial markets (Sornette, 2003), business recovery (Rose, 2007) and the vulnerability of small states to shifts in the economic system (Briguglio, Cordina, Farrugia & Vella, 2009).

Like other concepts discussed in this book, the concept of resilience has various meanings ascribed by various authors (cf. Klein, Nicholls & Thomalla, 2003). While some view resilience as an indicator of a system's capacity to return to its original state, others define resilience as a system's capacity to recover through an adaptation of its internal structure and processes. The concepts of hysteresis (this chapter) and time-irreversibility (the next chapter) reflect the logical impossibility of *returning* to a previous or original state; they imply that a system that recovers from the pressures exerted on it is somewhat different from what it originally was. Single-state resilience would mean inertia: a system that is unable to change would be as bad as a highly volatile system.

Consider, for example, the financial system as a complex adaptive system that is periodically prone to major changes, such as the stock market crashes of 1929 and 1987 (Sornette, 2003), and more recently in 2008. While its core functions, such as the allocation of resources, may survive such crises, other parts of it will change because they are deemed to be unsustainable in the new context. For example, regulators may decide that financial products, such as opaque credit default swaps, may pose too great a risk to the system and should be more tightly regulated. Thus, the financial system has a certain level of resilience that helps it cope with disturbances and shifts it from one state to another. However, since the new state of the system is not the same as the former state of the system (ibid.), it is important for us to focus on the system's ability to reorganize. Rose writes about the aftermath of crises in the public domain "[...] Policy-makers rushed to assure us of remedial actions to reduce the risk of future potential catastrophes. Where possible, they have emphasized preventative measures. [...] What is often overlooked is the fact that individuals, institutions, and communities have the ability to deflect, withstand, and rebound from serious shocks in terms of the course of their ordinary activities or through ingenuity and perseverance in the face of a crisis."(2007: 383)

Rose proposes a distinction between static and dynamic resilience to understand how social systems try to cope with crises. Static resilience refers to the ability to restore key functions quickly without extensive repair and reconstruction. For example, a public organization may be able to respond immediately to a new situation because it has a stock of public funds or the pertinent knowledge. Dynamic resilience, on the other hand, is relevant when a system needs more time to restore its core functions through reorganization and repairs. This introduces the dimension of time. This type of resilience is more complex because the system and the environment continue to evolve during the recovery phase, which means that the reorganization's aims should evolve too. This reorganization may take place at different levels. For example, individual public decision-makers may change their working routines and decision rules while the organization in which they work does not change much as a whole because others in the organization do not carry out these changes. Alternatively, a particular organization may adapt to the new situation whereas other organizations may not. On a broader scale, all of the organizations in a sector may reorganize, while this may not occur in another sector.

Thus, resilience in social systems refers to the ability to reorganize parts of the system or the system as a whole. This is often termed adaptation or adaptive behavior. Note that the system as a whole may change because of a possible non-linear relationship between what happens at the system level and activities by individual decision-makers. Also note that such resilience builds on the human capacity to anticipate and plan, implying that reorganization is more than just a passive reaction to disturbances. Reorganization is driven by the internal motivations of public decision-makers and the stimulus that comes from other decisions made by others (cf. Mileti, 1999). It may be possible to build up the capacity to deal with disturbances beforehand, but only to a limited extent. For example, an organization could build surplus stocks (e.g.

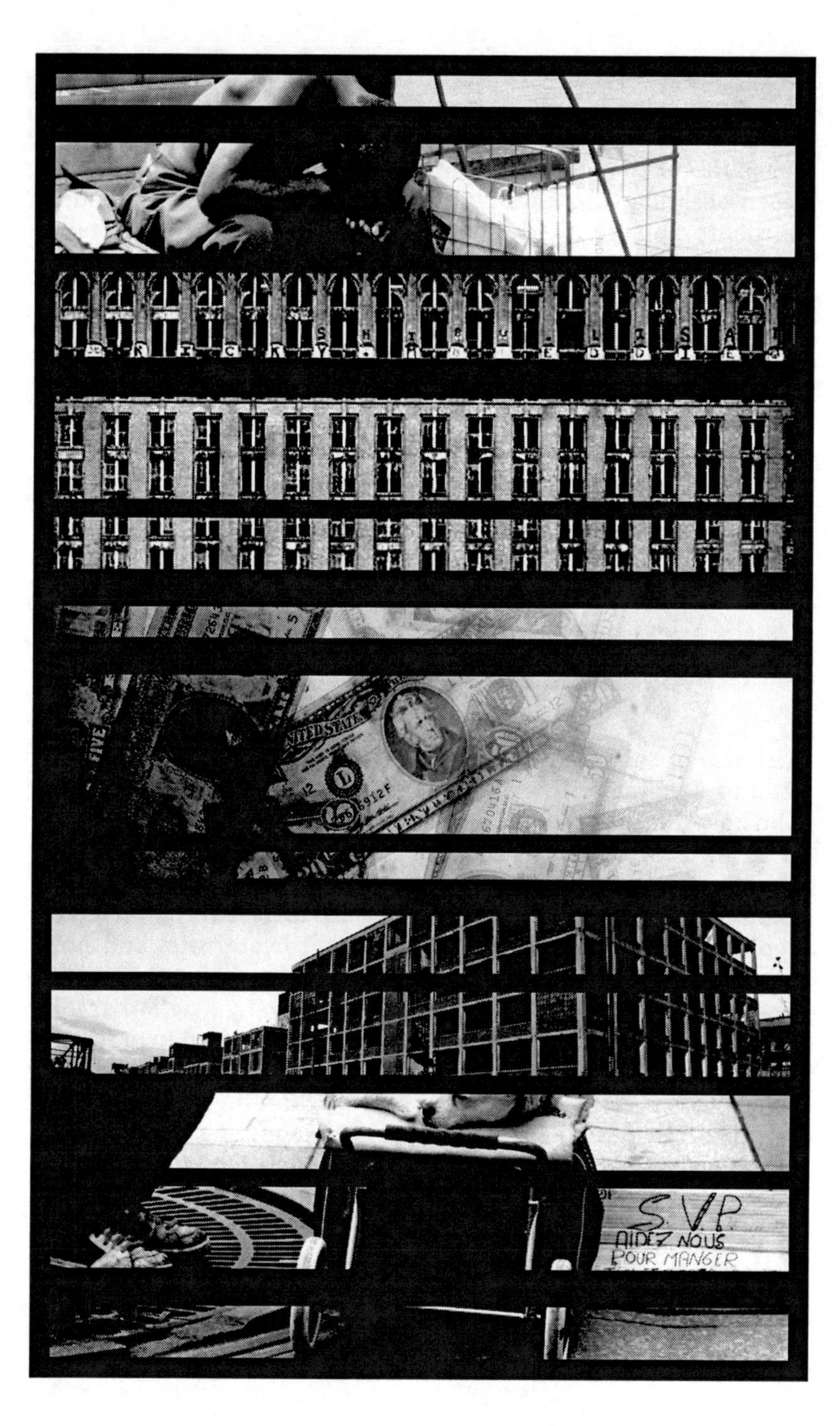
S.V.P.
AIDEZ NOUS
POUR MANGER

in the form of reserve funds or redundant staff), but building up this surplus is only part of the response and is potentially quite inefficient. Other factors that are relevant include the ingenuity of the individuals involved and their willingness to learn and adapt to changing circumstances- these are much harder to manage, if only because one does not know beforehand which qualities will be required. In addition, the disturbance itself may erode the resilience available. It may, for example, deplete the stocks or the ability or willingness to adapt to new situations. Carrying capacity and resilience are closely related concepts that explain how social systems cope with disturbances and both depend on the ability of individuals to interpret situations and deal with them accordingly. Thus, the next chapter will examine what individuals do when they are faced with complex situations.

3.4 CONCLUSIONS

The aim of this chapter was to identify the mechanisms that lead to dynamism in systems, demonstrate how they operate, and how this was related to public decision-making. Beginning with an overview of the basic mechanisms of feedback and chance events, the chapter then proceeded to examine how combinations of these could lead to different patterns of evolution (punctuated equilibriums and hysteresis) and how path dependency and lock-in influence the level of stability in a system. The extent of change possible in a system - whether it is able to continue in its present shape when subjected to pressure and resilience - is also affected by its carrying capacity, which refers to its ability to recover by reorganizing itself after it undergoes a period of turbulence. The challenge for (policy) researchers is to identify which feedback loops exacerbate particular pressures and how these impacts interact with a system's carrying capacity, causing it to change, or remain as it is, and if change does take place, how the system recovers.

A common mistake is to believe that a dynamic system is something that is continuously changing. This is not necessarily the case. 'Being dynamic' means that systems can be inert at times and change at other times. Perhaps the main challenge with dynamic systems is therefore not that they change or can be static, but the sequence of the two states that creates *uncertainty* for those who are part of them, such as public decision-makers. The same dynamics can determine the outcome when certain decisions are taken and policies implemented. As a senior civil servant from the Rotterdam municipal planning department remarked during the evaluation of their policies that were meant to discourage middle-class families from moving to the suburbs: "We like to think that we are in charge, but it sometimes seems that it doesn't really matter what we do." While policies do matter, the real world also intervenes with the good intentions and the comprehensive, synoptic plans that decision-makers attempt to develop. The next chapter focuses on how individuals cope with such complexity and the resulting uncertainty in decision-making.

CHAPTER 4.

BEING HUMAN: COPING WITH COMPLEXITY

4.1 INTRODUCTION

Complexity is not only a matter of systemic philosophy as laid out in Chapter 2, or a set of mechanisms as explained in Chapter 3. It is very much a part of the experience of people working in complex systems (cf. Wagenaar, 2005; Stacey, 2004). This chapter focuses on the experience of complexity and the ways in which people attempt to deal with the complexity they encounter when engaged in public decision-making. This is an important area as it is in this systemic complexity that politicians, administrators and public managers have to assess situations and arrive at decisions.

Research into how the mind operates when making decisions has been done in different ways. Although the approaches taken sometimes seem largely similar, they make vastly different assumptions and reflect distinctly diverse worldviews (Gigerenzer & Gaissmaier, 2011). Rational reasoning and attempts to improve decision-making by developing better decision-making methods are feasible when the subject is bounded unambiguously, and when all information is accessible. Such situations are dubbed 'small worlds' (Savage, 1954). While such environments are conceivable, namely in the shape of closed systems, they do not quite resemble the real social world where systems are open and perfect information is largely unattainable. The availability of information is an important difference between small worlds and the real world. In the words of Stiglitz (2010), a world where *near-perfect* information is available is very different from a world in which *perfect* information is fully available. And even if perfect information was available in the real world, humans do not have the capacity to estimate all possible options and their possible consequences (Simon, 1956), so the basic premise of models of rational decision-making will not be met (Gigerenzer & Gaissmaier, 2011). The previous chapters also argued that thinking in terms of complex systems subjects ideas about how such systems operate to multiple interpretations and requires diverging and intersecting boundary judgments.

An alternative approach, then, is to discard the assumption of small worlds as well as the premises of rational decision-making and focus more clearly on decision-making in the real world where people deploy different methods to make decisions. Rather than going through complicated calculations, they may use a simple and limited set of rules under which a large proportion of information is not even considered (Gigerenzer & Gaissmaier, 2011). For example, March's logic of appropriateness features three simple rules that act as a coping mechanism for people working in (public) organizations: what kind of person am I? What kind of situation is this? How is a person like me supposed to behave in such a situation? (1994). The answers to these questions help people to adapt to certain situations, and the questions can be applied in a conscious way. However, while we can make ourselves consciously aware of such behavioral rules and choose to adopt them, the human mind also operates a wide range of subconscious processes that are often as helpful or even more helpful in coping with complexity (Rilling & Sanfey, 2011). It is these processes that serve as the core of this chapter.

The purpose of this chapter is to explore the processes that come into play when people try to cope with the complexity of the real world. We argue that individual attempts to cope with complexity generate greater complexity. This argument is constructed in three steps, roughly based on the proposal of Peterson (1985). The first step is to try to understand how complexity is experienced by individual decision-makers (Section 4.2.1) This is done by discussing the nature of chance and time-asymmetry (Section 4.2.2), as well as of uncertainty and risk (Section 4.2.3). We argue that the simplification of complexity is inevitable for the sake of public decision-making (Section 4.2.4). To understand the process of simplification, the second part of this chapter looks at how complexity is processed within the human mind (Section 4.3.1). We focus in the subsequent section on the use of heuristics as mental shortcuts in decision-making (Section 4.3.2). We then argue that heuristics are both a means to cope with complexity, and a way to generate complexity in interactions with others (Section 4.3.3). The third part of the chapter (Section 4.4.1) introduces the concepts of autopoieses and self-organization (Section 4.4.2) to help explain the intricate workings of complexity, complexity reduction and group dynamics. The examples given throughout the chapter draw from the data gathered from respondents from a number of the author's research projects.

4.2.1 THE EXPERIENCE

Even though they have had significant experiences of complexity, it is quite likely that many of the people engaged in public decision-making have not bothered themselves with the complexity sciences. Such an experience comes in many shapes and guises and can even appear paradoxical. A respondent stated that he experienced complexity when too many variables had to be taken into account for a decision to be taken. At the same time, he acknowledged that limiting the number of variables also created complexity because of the things that were overlooked. What does this experience tell us about complexity? Our interviews allow us a provisional overview with five dimensions: [a] the perception of a great number of factors that are related in one way or the other; [b] confusion over those relations in terms of causes and consequences; [c] clashes of the goals, values and beliefs held by different actors; [d] time pressure 'to get things done'; and [e] the experience of sudden change in the environment of the decision-maker that can disrupt earlier plans and intentions. The recurring theme in these five dimensions is uncertainty over what may happen next, and it is this uncertainty that uniformly hinders the decision-maker in making his assessment and bringing about his aspirations.

While we would like life to be full of planned events that turn out in the way that is expected, the truth is that many things do not necessarily depend on intentional action and are very much rooted in chance. Recall the series adapted from Rescher as discussed in Chapter 1. 'Abababa' has repetitive markers that are not generally experienced as complex, whereas '*ca*a11*b*c763*cc*z' has some repetition, as well as other markers that seem more idiosyncratic and unpredictable. The same thing happens in real life. Some things are planned, some things are intended, and some things are favorable. Other things may be the opposite: unplanned, unintentional and unfavorable. But there are also many other possible combinations of attributes, including events that may be unintended but favorable. Cohen, March and Olson (1972) and Kingdon (1986), among others, recognized this about the way public decision-making processes develop. They recognized further that some decisions may take place only under certain favorable circumstances. In such situations, people may commit the fundamental attribution error, i.e. they reason that favorable outcomes are due to their own actions, whereas those perceived as unfavorable are due to the circumstances or actions of others (Ross, 1977; Ross & Nisbett, 1991; Sabini, Siepmann & Stein, 2001). In other words, we are prone to self-deceit and we create our own particular version of causality. The key point here is that getting the conditions right does not only depend on managerial activities or purposeful intentions, but very much on chance. Given the important role of chance in public decision-making, we need to look more closely at how chance impacts decision-making, how it creates uncertainty and what the consequences of such uncertainty are for human behavior.

4.2.2 CHANCE AND TIME-ASYMMETRY

Systems can be surprising simply because of the occurrence of random, unforeseen chance events (Sibeon, 1999), that have been described in Chapter 3. Chance can, and often is expressed in probabilistic equations. However, these are not helpful in increasing our understanding of the experience of such events, and their impact. In the Netherlands, the chances of high water levels leading to the breaching of dykes and consequent flooding is anywhere between once in 1250 and once in 10 000 years. The probability of flooding depends also on the height, strength and construction method of the dykes, which in turn depend on the value of the land and what the land behind the dykes are used for. Despite the low odds, flooding occurred first in 1993, and again in 1995 in roughly the same area, thus demonstrating the real differences between a probabilistic distribution and deterministic distribution. It also clearly illustrated the fundamental difference between distribution over a population and the individual experience. What mattered to people, especially those forced to evacuate twice in three years, was not the formal distribution of chance events, but rather the lack of regularity in their occurrence.

The crudest appearance of chance is when something happens that we cannot fully predict prior to its occurrence. This inability to predict is rooted in two dimensions, as explained by Rescher: "Natural impediments to prediction obtain insofar as the future is *developmentally open* – causally undetermined or underdetermined by the existing realities of the present and open to the contingencies of chance or choice. Cognitive impediments obtain insofar as the future is *epistemically inaccessible* to us, which may occur because we do not know operative laws (uncertainty) or the requisite data (pre*dic*tive myopia), or else because the interferences and calculations needed to obtain answers from laws and data involve complexities beyond the reach of our predictive capabilities (incapacity)" (1995: 41-42, italics original).

The first dimension of chance concerns the indeterministic nature of reality. A deterministic world is predictable because the relationships between its constituent elements are fixed, which means for example that a certain event *a* will always cause event *b*. The future of this world is fully determined by the current situation. An indeterministic world, in contrast, is one where certain events *may* or *may not* occur because of the conjunctions of conditions rather than the occurrence of deterministic causality. Some configurations may bring forth a more probable future than others, something we will discuss in more detail in Chapter 6, but the core of the argument is that the future is not fully determined by the present situation. Instead, it is very much developmentally open: "When appropriate initial conditions are given we can predict with certainty the future or 'retrodict' the past. Once instability [in the shape of events – LG] is included, this is not longer the case, and the meaning of the laws of nature changes radically, for they now express possibilities or probabilities."(Prigogine, 1997: 4).

One may argue that the occurrence of a certain event is not a matter of chance, but simply an indication of ignorance on the part of the observer. It might be that people perceive something as coincidental or unforeseen simply because of their failure to forecast correctly. This concern is echoed in Rescher's second dimension and the idea that we are unable to forecast because we haven't gathered enough data. Similarly, it could be because we can't understand the laws at work as the effort required to process all information is simply too high, or because the available information is evaluated selectively (cf. Beahrs, 1992). This presents us with an interesting question: if the future is epistemically inaccessible, how do we differentiate the causal role of our own inabilities from that of the nature of reality? Are we dealing with a rather large 'small world', or with an infinitely open and indeterministic one? It should be understood that the occurrence of chance in 'small worlds' is a very different thing than the occurrence of chance in the real world. And while there may be occasions or situations in the real world that meet the criteria of 'small worlds', public decision-making ultimately takes place in a capricious and indetermined world, as argued in Chapter 3 and as shown through the many examples in this book. While it is true that people go to great lengths to give meaning to the occurrence and effects of chance (cf. Alchian, 1950), it is equally true that time develops in a way that can only be partially traced to the initial conditions.

People see the inevitability of events following from previous events most clearly when they look backwards. History draws a straight arrow pointing at the current condition as the only logical consequence of sequences of events occurring prior. Looking into the future, however, the obvious sequences of the past are missing. People may build mental images of multiple futures ahead, some of which appear more likely than others, but they do so almost inevitably without foreseeing the chance events that can disrupt the predicted course of events. The arguments of time-asymmetry and time-irreversibility, as first introduced in Chapter 3 when we discussed punctuated equilibrium and hysteresis, mean that past sequences of events are not perfect predictors of the future. Time always moves forward and incorporates an element of chance. While the present may resemble something of the past, it is, in fact, another step into the future. Even a restoration of a previous situation (if possible) is not a return in time. Thus, people always face a certain degree of uncertainty.

4.2.3 UNCERTAINTY AND RISK

Chance, time-asymmetry and time-irreversibility turn the future into a collection of possible future states rather than a single definite state. Our predictive capacity is greatly limited, and we are therefore compromised in deciding which course of action is necessary to reach the desired goal or end-state. Adapting from Prigogine (1997) and Byrne (1998), we can (statistically) predict within a certain bandwidth how groups or ensembles of trajectories of social systems will behave. However, we cannot predict individual trajectories within the ensemble. In other words, the ensemble is not equivalent to the individual trajectory, but we need to understand the individual trajectory in order to understand which decisions to take in order to reach a desired end-state. In terms of decision-making, it becomes important to acknowledge that the individual route towards the best outcome cannot be identified. Alchian writes: "Under uncertainty, by definition, each action that may be chosen is identified with a distribution of potential outcomes, not with a unique outcome." (Alchian, 1950: 212). The reason for this is that the information available at $t=0$ gives probabilistic information, i.e. information about the set of possible outcomes, not deterministic information.

The experience of uncertainty is not just a matter of a difference between variability at the level of the population and the level of the individual. Other factors include our expectations about what will happen next, which are shaped by personal experiences and hopes (cf. Verzberger, 1995), the extent to which people feel that they can influence the present (cf. Alchian, 1950), and the extent to which they like the past (cf. Rescher, 1995). The latter is important because unwanted surprises tend to be more noticeable than desired ones. Thus, decision-makers are constantly faced with a certain degree of uncertainty that impacts their decisions (Elster, 2007). When is the right time to make a decision? Should I wait or act now? Do I need more information? What will happen if I carry out my decision? Decision-making with any degree of uncertainty therefore represents a jump into the unknown (cf. Flood, 1999). Consequently, the decision maker has to make a conscious or subconscious evaluation of whether to accept or avoid the risks associated with doing something, with the consequences being unknown to some extent.

The objectivist probabilistic approach holds that risk equals chance multiplied by impact. This is different from uncertainty because perfect information must be available to solve this equation. Although the objectivist probabilistic approach is helpful in the quantification of risk, it is not very helpful in assisting our understanding of what people experience when faced with uncertain outcomes. We therefore need to focus on the subjective perception of risk rather than the objectivist determination of it (cf. Renn, 1992; Vertzberger, 1995; Ellen & Gerrits, 2006). Confusingly, the subjective perception of risk bears closer resemblance to non-stochastic uncertainty than to the objectivist stochastic definition of risk. Stochastic uncertainty becomes concrete in the world of decision-makers when, for example, they try to determine the probability of something occurring as part of their assessment of whether a certain policy needs to be carried out, e.g. by developing a cost-benefit analysis or environmental impact assessment. From this perspective, risk can be determined because it can be calculated. An example of this is the calculations made to determine the level of risk in the construction of a new railway, as was the case in the projects mentioned in Chapter 1. While it is true that people involved in such projects tend to underestimate the calculable risks (Flyvbjerg, Holm and Buhl, 2002), in their defense, it is worth remembering that a considerable but indeterminable part of the risk involved in a project or program cannot be calculated at all. While the inherent risks may be made objective through risk indicators, they may not be experienced at the personal, or subjective level. As such, risk becomes subjective and cannot be expressed meaningfully in terms of risk distribution. Instead, each risk assessment becomes a highly individual consideration between the polarities of risk-avoidance and risk-acceptance (e.g. Loewenstein et al, 2001) and it is clear that the outcomes of such a subjective assessment exercise bears little relation to the outcomes of an objectivist assessment (Ayal & Zakay, 2000). Two dimensions define the subjectivist approach to risk: [1] the position of people relative to other people in the population; and [2] the personal, highly individualistic preference for dealing with uncertainty. The first dimension emphasizes the fact that people operate in groups, and that there are intricate relationships between the preferences of individuals and the preferences of the group as the sum of individual preferences. This means that the risk norm is a rather fluid, ambiguous norm that evolves with changes in individual preferences. Arguably, the term 'norm' can cause some confusion because a norm implies a stable situation, when risk preferences in a group are not stable in the long run. The second dimension emphasizes the fact that each individual has personal choices in how to approach uncertainty, as well as individual preferences as to how much risk is acceptable. This preference is not about the distance between 'real' risk and personal risks norms, but rather about the willingness to make a certain decision or undertake a certain action in an ambiguous situation. People tend to respond in very different ways to very similar situations. Some people may adopt a fatalistic approach and experience a feeling of powerlessness and consequently not undertake any action. Other people may opt for a more controlling approach and refer to standard operating procedures to 'solve' an ambiguous situation (cf. Ellen & Gerrits, 2007).

An example of how different people deal differently with uncertainty and risks in similar situations can be seen in interviews with public managers involved in the reorganization of a municipal urban planning department. One sub-department had run steadily for years, and under the guidance of its previous director, it had developed a standard routine for dealing with uncertainty and the risks involved when presenting well-developed plans to the city council that could accept or reject them. When the staff of the sub-department identified a risk (rejection of their plan by the council), they developed a routine that required each plan to be circulated among all directors for cross-checking. The plan was allowed to be forwarded to the city council only after it was approved by all the directors. The new director of this sub-department came from a different field, and confessed to knowing very little about urban planning. While she was exposed, arguably, to the same uncertainties and risks of having a proposal deferred by the council, she opted for a very different approach. Rather than demanding that each plan first gain her own approval, she allowed her staff to contact the council directly. She relied on her staff's experience of what the council was likely to accept or not, and empowered them to self-organize and correct each other as necessary. While the work of presenting plans and proposals to the city council had not changed, her different approach highlights how different people deal differently with the same risks and clearly demonstrates the interplay between personal and group preferences.

The argument above allows us to relate chance, uncertainty and risk for a given *population* to a developmentally open world (Figure 4.1). This world is characterized by sequences of events that follow from previous events. What is unknown, is whether this sequence of events is predictable or random. Deterministic sequences of events are predictable, and therefore tend to present *certainty*. However, among social systems, such certainty can only be found in small worlds, which are closed systems that are difficult but not complex, and where it is possible to obtain perfect information. Open systems, in contrast, are indeterministic because they are conjunctions of conditions. The possibility of chance events means that what happens next is uncertain. This *uncertainty* takes two forms: stochastic and non-stochastic. *Stochastic* sequences of events are non-deterministic sequences which can be assigned a certain *probability*. Although a number of possible future events are presented, the likelihood of the occurrence of a particular event can be determined without being fully predictable. Assigning a probability to the event allows us to come up with a *probability distribution* of all possible events. In contrast, *non-stochastic* sequences refer to the occurrence of future events whose likelihood cannot be determined in any way. That is, there are no indications of the probability of their occurrence.

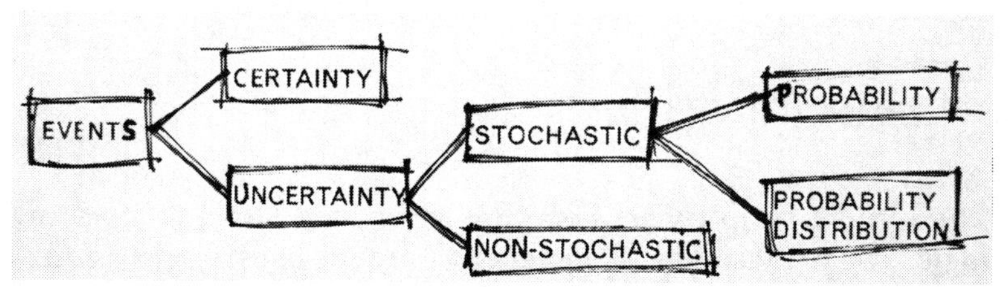

Figure 4.1: Dimensions of a developmentally open world. Courtesy of Peter Marks

The diagram above applies to populations. As we have seen before, what counts for a population does not necessarily count for an individual member of that population. Thus, the diagram changes meaning if it is applied to an individual who is making guesses about the future in order to determine what the best possible decision could be. When the subjective personal perspective of an individual is adopted, as opposed to the perspective of a population, these dimensions change their meaning. The differentiating factor is that the question is asked as to whether the individual *feels* certain or uncertain about the future consequences of a particular decision. Whether a certain choice brings about favorable results is the result not only of skill and insight, but also luck or fortuitous circumstances, one's standing relative to others, and one's ability to adapt to the circumstances one faces. The nature of uncertainty, chance and risk is that the venturesome could obtain better results than "logical, careful, fact-gathering individuals." (Alchian, 1950: 213) 'Risk' then becomes a defining category because individuals tend not to distinguish the probability of some event and its impact. People combine the two elements together in a subjective way, and this subjectivity is very important for understanding public decision-making (Jervis, 1992). Uncertainty may be experienced because the person has no clear idea how information should be evaluated. Conversely, certainty may be experienced simply because some information is not observed or considered and because decision-makers tend to create 'small worlds' by discarding information that they find distracting, irrelevant or troublesome. This process of simplification is paramount for explaining how individuals deal with complexity.

4.2.4 SIMPLIFICATION

The idea of simplifying complex situations has a compelling logic. Decision-makers generally crave order and sense-making, regularity, predictability and stability which would make their task easier and less prone to whimsy. Many public sector respondents to the interviews conducted by the author over the years expressed dislike in varying degrees for what they considered overly complex work and environments. When asked what they were craving, they mentioned a desire for simplification, such as a centralization of decision-making, clear and unambiguous rules, the convergence of different interests into the ´overall goal´, stable time-paths, and a powerful leader to unite them all. Scientists writing about decision-making in complex situations echo this desire and some have argued that a simplification of complexity is a good thing (e.g. Conner, 1998; Maguire & McKelvey, 1999; Harter, 2006; Solow & Szmerekovsky, 2006). However popular this argument is, it is important to remember that it remains largely irrelevant. Decision-making is not a choice between simplification and non–simplification as simplification occurs regardless of whether it is desirable, as we will discuss below. It is not a simple phenomenon by itself, and decision-making involves many forms of simplification, some of which are very subtle (e.g. Schwenk, 1984; Lefebure, 1999; Boisot, 2000; Cooksey, 2000). Simplification is thus both inevitable for survival, as well as (potentially) destructive if it results in the oversimplification of complex issues (Bearhs, 1992; Strand, 2002; Morcul, 2002; Teisman, 2005).

The temptation to simplify is rooted in the fact that decision-making in stable, knowable systems is much more straightforward than in dynamic systems. The series "abababa" is predictable ad infinitum and can therefore be approached with standardized routines. Apparent randomness, however, comes with uncertainty and requires more consideration than the mere application of a known routine. But while predictability and certainty (as opposed to unpredictability and uncertainty) are appreciated because they are easier to deal with and require fewer resources, decision-makers are not always in favor of simplification. In fact, they appreciate a combination of the simplistic and complex. Too much simplicity is disliked for its complete repetitiveness and predictability, while too much complexity is disliked for its ostensibly random nature and complete unpredictability. Too much predictability means also that there are no surprises, no changes, and no new outcomes. It is decidedly unchallenging. In contrast, a certain degree of unpredictability stimulates the brain. While the mental schemata called upon are already existing, the individual gets some joy from adapting those schemata to suit 'new-ish' situations that seem familiar yet have a significant degree of novelty (Berlyne, 1971). Complexity also allows for some ambiguity on the part of decision-makers. It frees them from making, for example, clear-cut but painful decisions.

Some of the respondents had been hired as trouble-shooters to deal with projects that had run aground for various reasons and in which standard approaches did not appear to be working. These respondents found an attractive degree of complexity in their work that made it interesting. Most commonly, they felt challenged by the ever-changing conditions, and did not generally complain about unpredictability, pure randomness or a lack of solutions. They did not generally feel their work to be a complete improvisation with unknown consequences. At the same time, most such respondents confessed that complexity did not sell very well amongst their stakeholders, and even though they appreciated complexity, simplification was what was largely desired.

The process of simplification takes many shapes. For example, Wagenaar (1997; 2005) analyses how people develop narratives to make sense of complexity. These narratives help to re-construct certain causalities whilst leaving out others. Sense-making through communication is thus one way of simplifying matters. However, we will follow the route proposed by Farnham (1990) and focus on the psychological dimensions of simplification for public decision-making by looking at how the mind generates and deploys mechanisms of simplification.

4.3.1 THE MIND

To this point, we have discussed the different ways in which complexity generates uncertainty about how the real world operates for public decision-makers. The processes of coping with complexity and the ensuing uncertainty are neurobiological in nature, and research into such processes helps clarify how decision-makers structure complexity and uncertainty and how semiotics are developed. Peterson (1985), among others, has researched how the mind functions in complex situations and how that influences public decision-making. Complexity generates information and that information reaches individuals through their sensory systems, after which it is processed and stored in the short-term memory and long-term memory, from where it can be retrieved. Peterson proposes that the sensory system and the processing and storage of information function as filters that select certain bits of information whilst discarding others in order to maintain structure and aid sense-making (1985).

The sensory system has five ways of receiving information from the environment: visual, aural, tactile, gustatory, and olfactory (Hunt, 1982, in Peterson, 1985). If a person pays attention, which is not a conscious process by definition, the information is received. If no attention is paid, the information is discarded within seconds. The interaction between the sensory system and the mind means that new information is added and evaluated to the existing structures of information. Information that is too different from existing information is more likely to be discarded. Conversely, information that is missing may be retrieved from memory by the mind to generate a coherent image of the world. This is such a seamless process that it creates the illusion of being a true representation of the real world, even though the information that has been added on may not be representative of the information that is missing. Over time, even the existing, older, information is reprocessed and will alter. The way the mind treats information gaps is a subconscious analysis of probabilities. It estimates what is needed to supplement the information it possesses for the coherent picture to emerge. The purpose of this information supplementation process may be that it enhances survivability (Sharkanksy, 2002; Newell, Paton, Hayes & Griffiths, 2010; Morçöl, 2011). Generating guesses instantly to fill gaps of missing information allows people to respond directly to situations rather than having to wait and process all the information available to construct a complete and balanced picture. This is important for decision-making because it means that decisions could be made quickly but also erroneously, because they were not based on actual, but rather supplemented, information.

Reaching a decision involves a search process across possible future decisions and their possible consequences. It may be tempting to think that human cognition is ill-suited to that task because of the way it handles missing information subconsciously. However, there are two reasons why this may not be the case. First, the results from that subconscious process are fed back into conscious thinking where they can be subject to conscious evaluation and consequent decision-making (Peterson, 1985). Secondly, there are indications that decisions based on cognitive subconscious shortcuts or cues, rather than rational or logical reasoning, may lead to good results. Gigerenzer and Gaissmaier (2011) demonstrate that such shortcuts performed on par or even better than other methods such as multiple linear regression analyses in a number of studies. In other words, there may be no trade-off between efficiency (of effort) and accuracy (arriving at the right decision). But before we discuss those shortcuts in more detail, we first need to return to the issue of rationality because it allows us to understand what 'better performance' and 'better decisions' are.

Epstein (1994) argued that decision-making can be broadly defined into two categories: (a) rational decision-making or a derivate form thereof, and (b) one "variously labeled intuitive, automatic, natural, nonverbal, narrative and experiential" (Slovic et al, 2007: 1334). While rational decision-making involves the processing of decision-trees, the intuitive variant supposedly can generate the same outcomes in a faster way instead of having to go through the calculations of each possible outcome and its consequences. Such shortcuts are a means of making decisions quickly, instead of having to go through the calculations of each possible outcome and its consequences (Tversky & Kahnemann, 1983; Gigerenzer & Todd, 1999). Notice how this reasoning takes bounded rationality and the impossibility of perfect information and consequent uncertainty as the point of departure (Tversky & Kahnemann, 1983). However, this type of reasoning appears also to be based on 'small worlds'. In an overview of the study of heuristics, Gigerenzer and Gaissmaier define heuristics as "a strategy that ignores part of the information, with the goal of making decisions more quickly, frugally and/or accurately than more complex methods" (2011: 474). This defines cognitive shortcuts as one way of weighing and forecasting, next to other methods such as Bayesian models. ´Small world´ reasoning enables cognitive shortcuts to be pitched against other methods to see which one delivers the best results. It means the researcher is able to determine what a good result entails, something that can only be established in 'small worlds'. It also allows performance to be studied, provided this is done in controlled experiments, as that approach caters for 'small worlds' because of the ability to strictly control environmental influences. However, one may question the extent to which the results obtained in such controlled circumstances could be relevant about the actions of decision-makers do in real, messy circumstances (Schwenk, 1984; Levitt & List, 2007; Rilling & Sanfey, 2011).

The issues mentioned here indicate two dichotomies in analyzing and understanding cognitive shortcuts. First, there is the difference between the use of such shortcuts as an alternative to other methods within the framework of bounded rationality for scanning probabilities and deploying fast and frugal decision-making (e.g. Todd & Gigerenzer, 2000; Gigerenzer & Goldstein, 1996) versus their use as a collection of non-rational behavioral guidelines for channeling complexity and assigning meaning to it (e.g. Nabi, 2003). The second dichotomy is between descriptive (which shortcuts do people deploy under certain conditions) and prescriptive (which shortcuts should be used when instead of other methods?) shortcuts (Gigerenzer & Gaissmaier, 2011). These two dichotomies help explain why, according to some researchers, decision-makers can arrive at different decisions, were it not for the interference of their own cognitive shortcomings or biases. The reasoning is that individuals could have gained greater utility without relying on their own cognitive cues (e.g. Gowda, 1999). However, as researchers of complexity, we are not interested in how things could have been, but rather how decisions are made in ambiguous environments. That requires finding out what cognitive shortcuts are and how they are used to cope with the vagaries of complexity. After all, even politicians and administrators are human decision-makers who have limited capacity to deal with the future possibilities of complex systems (Ghosal, 2001). A failure to appreciate the real limits of day-to-day political decision-making leads to unrealistic research instead of providing insight (Farnham, 1990). The question is: which shortcuts guide decision-making?

4.3.2 HEURISTICS

The term 'heuristics' is generally used to describe the group of cognitive shortcuts discussed above. Heuristics are intuitive, short and simple decision rules that achieve effort and time reduction by assessing a target property through a more readily available proxy (Kahneman & Frederick, 2002), and by examining and evaluating incomplete information (Shah & Oppenheimer, 2008). Heuristics can be deployed both consciously or can be activated subconsciously; it is very likely that many decision-makers use them without being aware that they are doing so. Heuristics are helpful in bridging gaps in information and in speeding up decision-making. They are environmentally-sensitive, i.e. they can be adapted to the local environment (Todd & Gigerenzer, 2000), and are relevant for understanding public decision-making processes (Miller, 2009).

We focus on descriptive non-rational heuristics, which come in several types (ibid.), such as search heuristics (for finding meaningful information), assessment heuristics (for evaluating information), and selection heuristics (for selecting options from the information that has been evaluated). These are broad categories and heuristics can be grouped in several different ways, because they form a chain between search, evaluation and selection. Still, these categories are used as a guiding thread to discuss some of the more prominent heuristics in decision-making.

Arguably, the heuristic that will be most easily recognized is the affect heuristic, which is used when individuals make a decision because they like the outcome of that decision or want to avoid some other outcome (Slovic, Finucane, Peters & MacGregor, 2007). The affect heuristic is rooted in the observation that emotion precedes rationality and that each consideration of a given situation and its possible outcomes is heavily influenced by such emotions (Zajonc, 1980; Shafir, Simonson & Tversky, 1993). Research by Damasio, Tranel, Damasio (1990) showed that patients with brain damage that impaired their ability to feel emotions whilst leaving their ability of logic reasoning intact were unable to arrive at decisions that would suit their own interest. Damasio argues this is because people use somatic markers in their decision-making, which are positive or negative markers based on past experiences that are attached to perceptions of future outcomes. The affect heuristic is based on "representations of objects and events in the people's minds" (Slovic et al, 2007: 1335), which are tagged to varying degrees with affect. In the process of making a judgment or a decision, people consult or refer to an 'affect pool', containing all the possible and negative tags consciously or unconsciously associated with the representation.

It should be stressed that the process of attaching affect to outcome perceptions when building up experience and the process of using the affect heuristics in decision-making is a subconscious process that can take place in tenths of seconds (Winkielman, Zajonc & Schwarz, 1997). This implies that people can be aware of their affective preferences but cannot control their interference in decision-making. Affect is determined by contextual information; that is, there must be something to compare a possible outcome with. It appears as if people find it more difficult to value certain options without contextual information, making the context a component for evaluation, i.e. there needs to be an environmental point of reference (Kahneman and Tversky, 1979). In public decision-making, this is often provided by the political, administrative and societal context. Another way of assigning value to an outcome is the effort heuristic, which is about the relationship between the effort that has gone into something and the perceived quality of the output. Generally speaking, people seem to value output that has taken more effort, in terms of time, money or otherwise, more highly than output that came about quickly, cheaply or easily (Kruger, Wirtz, Van Boven & Altermatt, 2004). This can help explain why policy programs or projects that have been through a long development time are difficult to kill, even if abandoning them is the right decision.

The availability heuristic holds that people base their preferences on whether they can assess the probability of outcomes by having a readily-available example (Tversky & Kahnemann, 1973; 1983; Miller, 2009), which is a clear example of assessment by proxy (Kahneman & Frederick, 2002). Slovic, Fischhof and Lichtenstein (1979) and Gowda (2001) give the example of more demands for increased regulation of airline safety rather than disease-prevention because airline accidents are more vivid and therefore easier to think of; they are more readily available in the mind than the consequences of insufficient disease prevention. Another aspect of the availability heuristic is that individuals are more likely to recall recent experiences to guide their decision-making than older experiences. Individuals are also more inclined to recall something they experienced themselves instead of something someone else told them (Miller, 2009). Related to this is the hindsight heuristic (Fischhof, 1975), which holds that people reconstruct the causality of events after they have happened in order to make sense out of it - something which Wagenaar refers to in the context of public decision-making (1997). The hindsight heuristic is important for understanding why history appears like a straight arrow, as argued in Section 4.2.1. It influences the way people reconstruct sequences of events. It was telling that two respondents in the Westerschelde case study corrected the case description regarding the sequence of various events. Not only did they have diverging interpretations, they were both doing this with data that stemmed from their own accounts that was recorded two years before.

Representativeness heuristics are geared towards singling out a specific option because of a certain instant association or perceived similarity connected to that option (Kahnemann and Tversky, 1972). The similarity heuristic (Read & Gruschka-Cockaybe, 2011) identifies an option by assessing its similarity to other known options. Rather than judging the probability of the option, it relies on recognizing how much the option looks like other options that the decision-maker is already familiar with. Many respondents explained that they managed public projects by recalling previous experiences that resembled the current situation and that they gained a sense of future directions by assuming that the current situation was similar to the previous situations in which they had been engaged. The recognition heuristic (Hilbig & Pohl, 2008) extends this by asserting that if one option out of two is recognized, decision-makers are likely to choose the one they recognize over the one they do not (Goldstein & Gigerenzer, 1999). The choice is a binary one: either the known option is liked or disliked, which connects it to the affection heuristic. Such heuristics could explain why repetition of the same or similar policies or projects are not uncommon in political decision-making.

Although the heuristics mentioned above are environmentally sensitive, they are not necessarily related to the direct social exchanges of the decision-maker. But it goes without saying that there can be heuristics based on the fact that people are social beings who connect with other people (Marsh, 2002). For example, Yamagishi, Teral, Kiyonari, Mifune and Kanazwa (2007) concluded that outcomes of games such as the prisoner's dilemma cannot be explained without understanding that people deploy social heuristics based on reciprocity, such as the matching heuristic (mirroring the choices of others) or the control heuristic (believing that one's choice determines other's choices). Such heuristics focus on social exchanges between people (cf. Kiyonari, Tanida & Yamagishi, 2000). Other forms of social heuristics include social comparison and social imitation, grouped under the header of social affiliation. Social comparison evaluates people and oneself against a set target, e.g. in terms of performance. As such, it can work as an adaptation technique in group dynamics. Social imitation helps people copy existing routines, speech, and behavior (Marsh, 2002). These heuristics are important for establishing and maintaining group strength.

One group of heuristics that is of particular importance to political decision-making is the use of moral heuristics (Sunstein, 2005). Public decision-making involves normative stances so it is only natural that moral cues direct what people decide to do (ibid.). However, attention on moral heuristics is relatively recent. This is possibly because, unlike other types of heuristics, it is impossible to prove with moral heuristics whether the application of a certain heuristic leads to seemingly erroneous results. A moral judgment makes demarcations between 'right' and 'wrong'; instead of revealing errors in decisions, it indicates whether the individual making a certain judgment values things differently than the researcher or observer. Indeed, the right or correct decision can be an entirely different thing than estimating the distribution of options and pay-offs. Koehler & Gershoff (2003) describe an experiment in which people preferred a higher chance of fatal accidents from car crashes over a lower chance of death because of malfunctioning airbags. Sunstein (2005) argues that a betrayal heuristic is at work here - people are willing to accept a higher risk of death in comparison to the chance of being betrayed by a company that makes devices to protect them from the same risk of death. While this may be true, it is possible that people make a choice because they think it is morally good or wrong. Choices are not mere technicalities that can be executed correctly or incorrectly, and ethical judgments are an important part of the public debate. Other moral heuristics include 'playing God', which provides cues as to how far governments should intervene in nature, including human nature; and the outrage heuristic, which gives cues about the magnitude of punishment following human deviations from societal norms, such as crime punishment.

The heuristics described above are deployed when searching, assessing and selecting among options. Gigerenzer & Todd (1999) propose a fourth category: heuristics that end the search, evaluation and selection. Arguably, the most prominent one is Simon's (1955) satisficing heuristic (Bendor, Kumar & Siegel, 2009): the process stops when the decision-maker feels satisfied with the process. Satisfaction could be met when e.g. the first option that leads to an exit is presented (Martignon et al. 2003). An alternative is the hiatus heuristic (Gigerenzer & Gaissmaier), in which the process stops when the most available option, usually the most recent one, is retrieved. It could be argued that many of the heuristics are combinations of the search, evaluation, selection and stopping processes, especially since many of them are intertwined together when they are deployed or used. They are powerful means for coping with complex systems (Ghosal, 2001; Sunstein, 2005). However, heuristics are not only a response to complexity but can also be a source of complexity. To understand the latter, we need to see how simple rules can potentially create complex behavior at the group level.

4.3.3 SIMPLE RULES, COMPLEX BEHAVIOR

While heuristics are simple behavioral rules that present a way of coping with complexity, they themselves are a source of complexity. Scientists such as Holland (1995), Langton (1986) and Reynolds (1987) have experimented with computational models in which programmed singular units were able to create complex patterns of interactive behavior out of a limited set of simple behavioral rules. Reynolds' computational simulation of the so-called 'boids' (1987) serves as an example (e.g. Smith & Stevens, 1996). This simulation aims to recreate the flocking behavior of birds or fishes. Instead of trying to program all the overall properties of flocks, which would require a lot of complicated code, Reynolds programmed three simple behavioral rules for each individual boid: to steer towards the average heading of the other boids, to steer towards the average position of the others, and to avoid crowding in small clusters. These individual behavioral rules produced swarm-like patterns and the swarm even managed to avoid static objects in the simulation whilst maintaining formation – something it was not explicitly programmed to do. Thus, the flock did not only emerge and maintain itself, it also showed adaptive capacity in being able to adjust itself to the circumstances encountered in its movement.

This model and similar others have sparked a great interest in the study of the emergence of complex and adaptive group or system level behavior from limited sets of simple rules deployed by individual actors. It has provided impetus also for the emergence of a body of research that uses agent-based modeling in the way discussed above to inform public decision-makers about the possible consequences of their choices (e.g. Kim, & Lee, 2007; Johnston, Kim, & Ayyangar, 2007; Koliba, Zia & Lee, 2011; Desouza & Lin, 2011). One such series of studies focusing on the dynamics of the process of decision-making in groups is that by Axelrod (1984; 1997; 2006). The basic assumption made in his and other such models and theories is that complex interactions and situations can result from the individual following simple rules as described in the previous section. For example, suppose one actor in a governing coalition needs to gain support for a policy proposal. A popular political rule of thumb dictates that the actor offer something in return to the other members of the coalition as the fastest way to gain the required support. It is a simple rule, that over time, structures human behaviors and creates intricate patterns of mutual support and promises. To achieve something new within that particular coalition requires a disentangling of the patterns in order to find a policy window. However, decision-makers don't apply such rules sequentially, but in many different combinations, and as we saw in the previous sections, such rules are also context-dependent.

The idea that such simple, local, and actor-bound, behavioral rules can generate complex behavioral patterns at the group level without design or superimposed control is summarized under the label of 'emergence'. Emergence can be defined as "the arising of novel and coherent structures, patterns and properties during the process of self-organization in complex systems" (Goldstein, 1999: 49). Self-organization in this definition refers to the fact that there is no top-down coordination that leads to the establishment of particular structures, patterns and properties. Instead, they arise from local interactions in the most literal sense of the word. Goldstein then defines five discerning properties of emergence (ibid.), of which three will be summarized here because they show how emergence corresponds to the properties of systems discussed in Chapter 2. First, the properties of a complex emerging system are novel in so far that they are non-linearly related to the properties of the rules from which the systems emerges. In other words: the properties of the rules don't allow an accurate prediction of what the emerging system will look like. Second, emerging systems maintain coherence over time, thus being somewhat permanent. However, this doesn't mean that such systems are static. This leads us to the third property of emergence, which is that it is very much dynamic. Because the analysis of heuristics and the thesis of emergence can help us to further understand how rules bring forward complexity, in the third and final part of this chapter, we discuss the relationship between behavioral rules and the emergence of social systems.

4.4.1 THE GROUP

The argument so far has been that decision-makers experience enduring complexity and uncertainty to which they respond with simplification through the use of heuristics and schemata. It was then argued that such heuristics or simple rules are both a response to complexity and themselves a source of complexity. They help structure the complexity of systems in which decision-makers find themselves in, and at the same time, generate complex systems. This is summarized in the concept of emergence. Emergence helps bind together the different strands in this chapter because it relates the actions and considerations of individuals to the dynamic at the group level in a non-linear fashion. We want to focus here on a specific type of emergence, namely self-organization (cf. Wolf & Horrevoet, 2004) as the structuring principle that controls the interaction between individual decision-makers and the behavior of groups. It is useful thus to revisit the work of Luhmann, who was first introduced in Chapter 2. His work on social systems (e.g. 1984; 1995) helps us to understand the relationship between the individual and the group, as he sees social systems as both the means of coping with complexity and the outcome of such complexity (cf. Beahrs, 1992). In his view, social systems create "islands of lesser complexity" (Luhmann, 1970: 116; Laermans, 1996) where unstructured complexity is converted to structured complexity (cf. Murray, 2003). His central thesis is that social systems consist of groups of individual actors (cf. Seidl & Becker, 2005) who aim to maintain themselves in terms of structure and property in a complex, changing environment (Luhmann, 1977; Bednarz, 1984). From this perspective, social systems act as a selection process that filters information for its constituent actors whilst at the same time establishing itself through that selection process. The selection process consists of different intersecting mechanisms: boundary judgments, communication, reproduction and semiotics. The best way to approach these mechanisms is to understand that they are not separate from each other but rather different dimensions of one overall property of social systems called *autopoiesis*. The following section discusses autopoiesis as a specific form of social self-organization.

4.4.2 AUTOPOIESIS AND SELF-ORGANIZATION

We argued in Chapter 2 that social systems are open by definition. Social systems therefore exist by the ability of their constituent members, such as public decision-makers, to make demarcations between what they consider a system and what they see as its environment. This relationship between the system and the environment forms a point of reference against which the system's members assess and select new information. But exactly how are the boundaries of social systems established and how are they maintained? Luhmann, discussed earlier in Section 2.3.2, took his cues from the biological theories of system reproduction as first introduced by Varela and Maturana (1974), better known as autopoiesis. Through the lens of autopoiesis, a social system emerges through established patterns of interactions and communications between people, and it regenerates because certain interactions and communications are preferred above other interactions (Blom & Haas, 1996). This process is recursive, meaning that the results of earlier interactions and communications form the basis for further interactions and communications, i.e. a replication of the past. For example, when people go into an informal or formal meeting, their conversations are usually built on what has been discussed during previous meetings or in other interactions outside of the meeting room. The conversation rarely starts on a blank sheet (Laermans, 1996). The act of communication is therefore an act of reproducing the system, and social systems are networks of consecutive communications. This recursive pattern establishes system boundaries over time and secures autonomy because they determine what is inside the system and what is in the environment. The key to reproduction is then that elements of the system refer to other elements that mirror the properties of the same elements – the system is thus self-reproductive or autopoietic.

Such social systems depend on a degree of closeness between actors, because they require stability or maintenance of the status-quo. However, they are not so closed to extent that they do not need an environment. On the contrary, it is the constant confrontation with the environment that reinforces the reproduction of the system as people experience what sets them as a group apart from the environment, specifically, what sets them apart from other people and groups. Incentives from the environment are assessed against the system's own nature and workings and are consequently selected when they confirm the system's being. As such, the system builds its own future, because its members are inclined towards choosing future possibilities that reinforce the current state of the system. Because each system iteration is not radically different from previous states, the systems function as islands of lesser complexity. This reduction in complexity comes about because the selection-and-self-referential process structures complexity in a way that is familiar to the actors in the system. Incentives from outside that are experienced negatively or regarded as the proverbial 'white noise' in fact help reinforce the system by establishing the difference between 'us' and 'them'.

Communication, of course, is not static. The argument above does not imply that communication is nothing more than repetitions of an existing repertoire. Communication is a process and therefore has a temporal aspect (Laermans, 1996). Each time people engage in communication, it is possible to either continue the previous communication, or introduce a new subject, i.e. to alter the content. Each occurrence presents a choice between continuation and change. It requires that people assess their communications and understand when and how to interfere with the existing theme. Luhmann´s view is that people can never fully comprehend what other people with whom they interact think about and how they experience complexity. Instead, people simply base the information they send and receive on their expectations of what others perceive. As Bednarz says: "the complexity of the social world is constituted by my ability to assume the perspectives of others. [...] The assumption of the perspectives of others, however, exacts a price: their unreliability." (1985: 64) If people cannot fully comprehend the experience of others, expectations about what the other experience become the basis of communications - much of which is rooted in past experiences, as argued in Section 4.3. Public decision-making is no exception to this (e.g. Monroe and Epperson, 1994).

It is in this context that heuristics help social systems to achieve and maintain their stability. Luhmann acknowledges that people use only a limited set of rules to assess communication and calibrate their expectations. Both the assessment and calibration are geared towards an expectation of the repetition of past events, routines, behaviors and so on – and are therefore essentially conservative. The processing mechanisms are inclined towards conserving the existing order in that the structure of information that does not comply with the standards is usually reinterpreted, ignored or discarded. This is especially the case in public decision-making (Brans & Rossbach, 1996).

The argument of autopoiesis above means that social systems are capable of producing and reproducing themselves, and of developing structures and properties without superimposed, 'outside' steering or deliberate design. A more general term for this is self-organization. A working definition of self-organization that summarizes the above is provided by Wolf & Holvoet: "Self-organisation is a dynamical and adaptive process where systems acquire and maintain structure themselves, without external control." (2004: 7). The concept of self-organization is used in many disciplines to describe similar phenomena, ranging from cybernetics (e.g. De Beer, 1981) to urban planning (e.g. Portugali, 1998). Therefore, while autopoiesis is an important form of social self-organization, not all self-organization is autopoietic, such as in the case of urban systems or traffic systems (e.g. Marshall, 2009; Bertolini, 2010). De Wolf & Holvoet put emphasis on the systemic properties leading to the establishment of stability, specifically: autonomy from the environment, resilience against external shocks and increasing order over time. For the record, self-organization is sometimes seen in the literature on organization theory as something that can be created or designed with the intent of opposing hierarchies (cf. Stacey, 2005). Here we consider autopoiesis or self-organization as something that is inherent to human nature, and therefore as something that takes place regardless of whether it is deemed desirable or not.

An example from the Unterelbe case-study already presented in Chapter 3 may help shed light on the self-propelling and self-organizing mechanisms of autopoiesis. For decades, the development of the port was determined by a relatively closed group of actors consisting of administrators and senators of the state of Hamburg, its departments such as the Department for Economy and Labor, several federal organizations such as the Waterway and Shipping Direction Nord, the Hamburg Port Authorities and related organizations such as the local Chamber of Commerce, and Hamburg Hafen und Logistik AG, the main operator of the port. The dominant discourse was that the port was inherently good for the whole region and therefore the extension of the port was not to be disputed. The size, dominance and uniformity of the group of actors in charge meant that the subject was, in fact, never disputed.

The unchallenged growth of the port in all its dimensions came under attack during the mid-1990s. Fear that a further deepening of the estuarine channels would harm the environment, damage the river embankments, increase the risk of flooding, and hinder fisheries and recreational shipping led to increasing public protests and a strengthening of the position of pressure groups. This did not at first impact the decision-makers who generally downplayed the risks and pointed to the adherence to appropriate legal procedures when confronted with opposition. However, what was generally unknown was that the organization issuing the building and dredging permits to the port authorities was the same one tasked to assess (and reject) the appeals lodged against these same permits.

The change of the estuary, as discussed in Chapter 3, and the subsequent social protests placed severe pressure on the unity of the decision-makers. Some members of the decision-making bodies started to question the status-quo which dictated never-ending and largely unhindered port extensions, and this eventually resulted in the appointment of a mediator. Together with the many external opponents, the mediator began to work on forging a long-term vision for the region as a whole that then allowed for a focus on the role of the port within that region. The development of the vision in itself helped channel new ideas into the decision-making, and new possibilities that were not previously considered became a key part of the communication.

The example above demonstrates the workings of autopoiesis in a nutshell. In the first stage, a group of actors come together and are held together by a powerful singular argument that the port should be extended simply because it had done well in the past, and was continuing to do so. This dominant view drowned out all others, and the group's existence and reproduction was established *without the actors being aware* of their singularity. This is an important point because it helps explain why, in the second stage, interference and increasing complexity from the outside only reinforced the decision makers' persistence. By maintaining internal communication and rejecting communication with opposing voices, they remained firm in their belief that they were acting in the right way. After all, everyone they talked to agreed with them. It explains also how it came to be that the issuer of permits was allowed to assess the appeals. In the third stage, greater pressure led to significant change that was triggered first by the appointment of a mediator, but gradually also by the opening of alternative futures through the visioning exercise. Autopoiesis and self-organization are powerful explanatory mechanisms with which the emergence, conservation and deconstruction of social systems as a means of coping with complexity can be understood. The observation that complexity necessitates simplicity, in turn generating complexity means that the story can now come full circle.

4.5 CONCLUSIONS

We set out to explore how the complexity of a developmentally real world is experienced at the level of individual decision-makers, and found out that simplification is an important process with which to cope with that complexity. Simplification occurs in the sensory system, in the short and long term memory and through mental schemata - where information is omitted, selected and supplemented in order to create a coherent picture. Special attention was paid to the working of heuristics as a set of simple rules that help us arrive at decisions in complex systems without having to go through and calculate the implications of all possible decision outcomes. We then introduced the concept of emergence and argued that such simple rules are generators of complex patterns of interaction between actors. Theories of autopoiesis and self-organization help us to understand how social systems emerge as both a *means* and a *result* of coping with complexity. They thus generate complexity and further contribute to the experience of complexity. Figure 4.2 summarizes the arguments of this chapter.

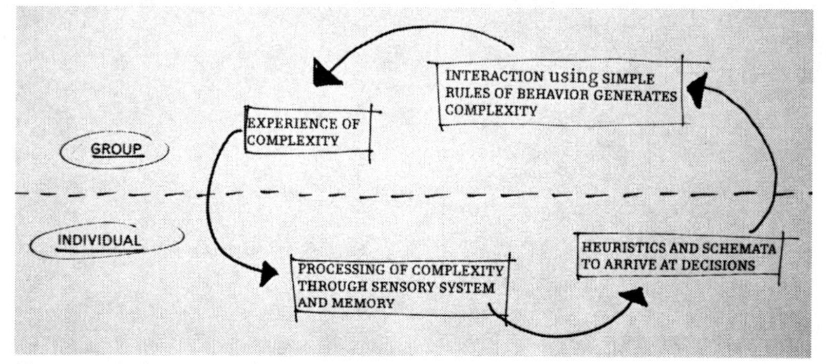

Figure 4.2: how people experience, process and generate complexity in their interactions with others

We have now followed three major strands in our quest to name, analyze and comprehend the complexity of public decision-making. In Chapter 2, we looked at the nature of complex systems. In Chapter 3, we looked at the dynamics of such systems and in the current chapter, we examined the role of the individual decision-maker coping with complexity in his daily interactions. Note that decision-makers are part of the systems they operate in and are crucial in generating complexity. Since public decision-making takes place in multi-actor settings where many actors have competing, conflicting, diverging or converging expectations and behavioral rules, it is an understatement to say that achieving something in the public realm is a dynamic challenge. We may now start to appreciate why public decision-making can be such a volatile undertaking. The next step is to integrate the three strands into a coherent framework, and it is this task that we will undertake in the next chapter.

CHAPTER 5.

COEVOLUTIONARY PUBLIC DECISION-MAKING

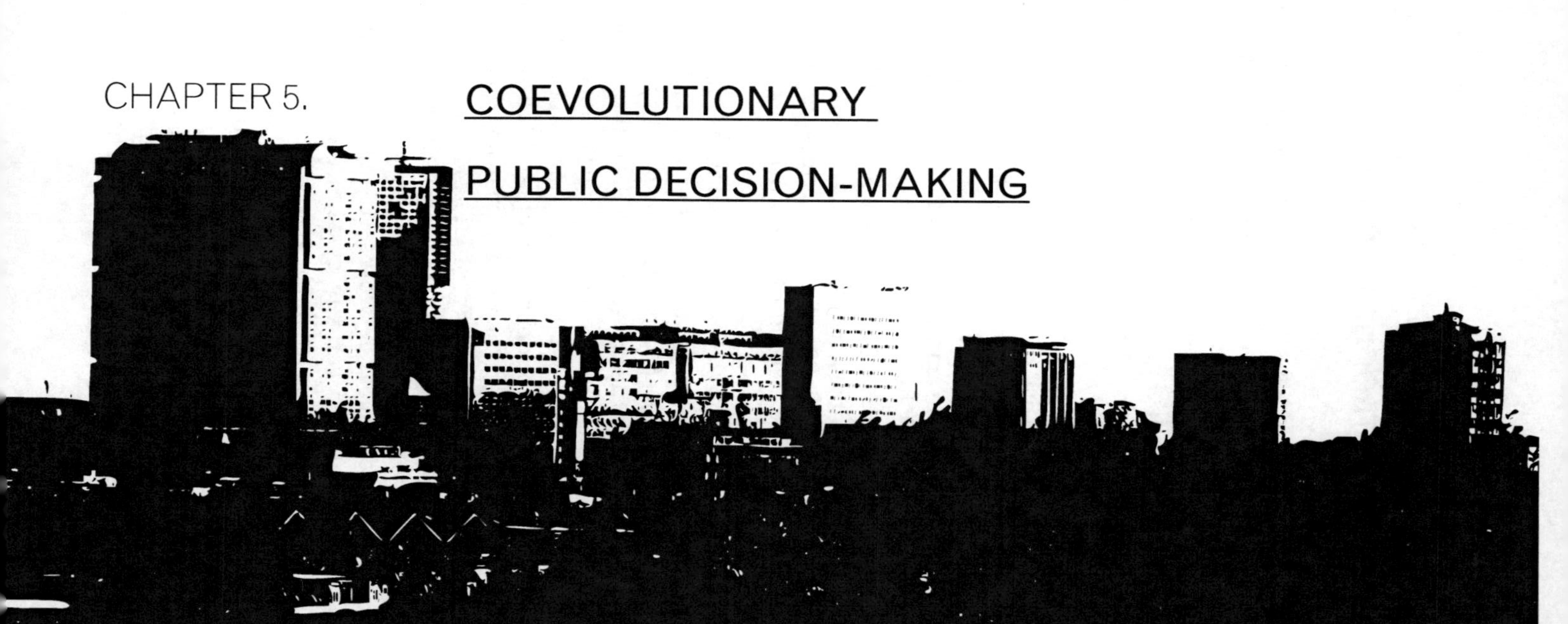

5.1
INTRODUCTION

The previous chapters explored the complexity of public decision-making, focusing on its systemic and dynamic nature and the role of the human mind in coping with complexity. This chapter presents a complexity-informed conceptualization of public decision-making by connecting those three dimensions. Therefore, we begin by briefly summarizing the findings of the previous chapters.

The first chapter used a thought-experiment to explore descriptive complexity. It showed that the world in which public decision-making takes place is an erratic one, where recurring and recognizable patterns forming particular configurations in time and space interact with idiosyncratic events. The resulting uncertainty about the consequences of these interactions makes it difficult to deal with them, even when the degree of uncertainty is relatively small. Complexity as described in the first chapter is driven by the systemic and dynamic nature of the world, a combination that generates novelty.

The first step in understanding this situated approach to reality was to discuss its systemic foundations in more detail. We explored the different iterations of systems thinking and systems theory, from Von Bertallanfy's general systems theory, Parson's functionalism, Checkland's focus on communications and meaning, to Flood's attempts to integrate these ideas under the banner of complex adaptive systems. Along the way, we included other complicating dimensions, such as semiotics and the blurring of hierarchies, and also reconnected the narrative with important notions from the domain of public administration and public policy, in particular, the idea of governance networks. Using research on the workings of the Dutch railway system, we argued that complex adaptive systems could be a basis for conceptualizing policy action systems. Such systems oscillate between homogeneity and heterogeneity because of environmental pressures and the consequent changes in the permeability of their boundaries.

The main argument of the second chapter is that the world's dynamism can be explained by feedback loops and non-ergodic chance events. Their cumulative impact leads to four dynamic effects. Punctuated equilibrium and hysteresis focus on system transitions, emphasizing that systems may not evolve gradually and restoring them to something resembling their previous state after they have changed may be difficult. Path-dependency, increasing returns and the consequent lock-in explain the emergence of inertia and the lasting impact of initial conditions on system trajectories. We enhanced our understanding of system dynamics further by introducing another property of systems, i.e. their carrying capacity. This refers to the extent to which a system is able to withstand and respond to disturbances through static and dynamic resilience. Together, these mechanisms and effects help explain how a routine policy decision to deepen the Unterelbe estuary in Germany led to a shift in the natural system that was unforeseen, unexpected and undesired. We can now explain how that could have happened and why it turned out to be rather difficult to reverse that unwanted situation, providing a real-life example of how non-linearity and time-irreversibility work in practice.

The third dimension of complexity is the way individuals receive and process information and their subsequent actions. We shifted our lens in the fourth chapter away from the system level to the level of individual decision-makers and their experience of complexity. It helped us to appreciate the nuanced differences between different kinds of subjective uncertainty and how time-irreversibility impacts those experiences. The central argument in this chapter was that complexity is filtered, structured and simplified so that it remains comprehensible for making certain decisions. If the world is developmentally open, that is, there are various degrees of uncertainty about the future, decision-making is affected because it is not entirely certain what that decision's results will be. Decision-makers cannot compute all the possibilities in a rational way. Instead, they compress complexity and use decision heuristics as guidance for making decisions. Such heuristics not only help in coping with complexity, but also serve as a drivers for self-organization and, perhaps paradoxically, to create more complexity because of the adaptive moves of decision-makers. Self-organization and autopoiesis are a further deepening of our understanding of homogeneous and heterogeneous policy action systems in response to a dynamic world.

The narratives from the previous chapters left us with

a number of related strands that we need to tie together in this chapter: a narrative about structures (Chapter 2), a narrative about processes (Chapter 3), and a narrative about actors (Chapter 4). The intersection of structure and process is marked by *temporality* as this is where integrity of systems in terms of connections and the feedback that flows through the connections are established and re-established in response to incentives. *Adaptation* marks the intersection of process and actors, where dynamic processes change the fit between decision-makers and environment, and where the former attempt to (re-)gain some level of fit with the dynamic environment. The intersection between actors and structure is marked by *diversity*, as actors self-organize in heterogeneous and homogeneous policy actions systems, i.e. when they cluster in certain groups to reach their goals.

These three identifiers will be used later in this chapter to structure the aspects of dynamics for understanding the complexity of public decision-making. Public decision-making can be characterized as a move forward in time without a clear end-state or final goal because of the influence exerted by the dynamic environment that prevent it from reaching such an end-state. But decision-makers are not just floating passively along with the stream of events that happen to them. Their considerations and actions matter and exert influence on the environment. In short, we may regard the relationship between public decision-making and the environment as mutually influential or *reciprocal*. A more specific term for this pattern of reciprocity is *coevolution*. We will combine the strands from the previous chapters to develop a coevolutionary revision of public decision-making processes.

The general idea of coevolution, its background, and the argument as to why it matters in public decision-making is presented in Section 5.2.1. Moving beyond the broad logic of coevolution, we operationalize its main mechanisms - variation, selection, and selection pressures - and use them to analyze public decision-making (Section 5.2.2). Fitness landscapes are a heuristic device to illustrate, demonstrate and discuss coevolutionary dynamics. The general idea of fitness landscapes and their value for understanding coevolution is introduced in Section 5.3.1, before our particular model of such a fitness landscape is shown (Section 5.3.2). We then use that fitness landscape to demonstrate a number of dynamic features, namely: adaptation, diversity and temporality as the basic conditions (Section 5.3.3), ruggedness and adaptive walks (Section 5.3.4), and attractors and the trajectories of systems (Section 5.3.5). The chapter closes with a condensed narrative that integrates the previous chapters to present a coevolutionary revision of public decision-making (Section 5.4).

5.2.1 COEVOLUTION IN DECISION-MAKING

Some authors have argued that evolutionary mechanisms are useful analytical concepts for understanding the complexity of the structures and processes public decision-makers, and the researchers who study them, are faced with (Sementelli, 2007). According to Kerr, such mechanisms "[...] can provide us with a more versatile, sophisticated and complex explanatory framework than many of the theoretical tools which we currently have at our disposal." (2002: 355) Daneke (1990) outlined a research agenda that framed public decision-making in complex systemic and evolutionary terms, building on the work of Corning (1983) and others. However, evolutionary approaches to public decision-making have not developed much since then (Kerr, 2002; Witt, 2003; Bergh & Kallis, 2009). Kaufman's account (1991) is one of the more well-known ones. It pays explicit attention to the limits of decision-makers in exerting influence on their environment and in reaching the desired outcomes. Those outcomes, Kaufman says, are the result of a partly random conjunction of several factors, which coincides with our findings from the previous chapters.

Although Kaufman uses the vocabulary of evolution to describe a real phenomenon, he does not use the concepts as explanatory mechanisms. While much theorising in the social sciences has an implicit or explicit evolutionist or evolutionary character (Sanderson, 1990; Nelson, 2006), there is a difference between discussing long-term incremental change in social systems in general and analyzing what causes those changes. This discrepancy highlights the different interpretations of the use of evolutionary theories or mechanisms in understanding social complexity (cf. Silva & Teixeira, 2009). A principal distinction is between the logic of evolution and the mechanisms of evolution (cf. Porter, 2006). Accounts that fall into the first category consider change as being evolutionary, i.e. incremental and without a clear end-state, while studies in the second category move one step further and dissect those incremental changes by deploying a specific set of explanatory mechanisms, such as variation, selection, and retention, which we flesh out in further detail in Section 5.5.2.

So far, we have used the notion of evolution in a generic way. However, a reciprocal approach to evolution means that evolution becomes *co*evolution. This concept is rooted in evolutionary biology and was first coined by Ehrlich and Raven (1964), who observed that groups of organisms evolved through reciprocal selective interaction (Norgaard, 1984; Odum, 1971). An organism's mutation can be explained by looking at the selection pressures on that organism from the environment. Coevolution explains that the mutation of an organism in turn affects the environment of that organism. The explanatory power of coevolution is situated in the patterns of mutual or reciprocal influence that arise between organisms, or, in the context of this book, between decision-making and the environment.

An explicit attempt to understand the coevolution between systems and the role of decision-making in that coevolution can be attributed to Richard Norgaard (1984; 1994; 1995; Gual & Norgaard, 2010; Kallis & Norgaard, 2010). He focuses on the coevolution of social and physical systems and argues that over the years, people have engaged in patterns of feedback loops with physical systems, using them to suit their needs, such as in agriculture (Norgaard, 1994; Moreno-Peñaranda & Kallis, 2010). In order to deal with the consequent feedback from physical systems and to optimize their use of that system, they are pushed to create increasingly individualized task specifications and more complex institutions. In other words, social systems try to change physical systems for their own needs, but the results of that pressure lead to physical changes, which in turn lead to a response from the social system, which the physical system then responds to with yet another set of changes, and so on. Over time, the complexity of this pattern of reciprocal influence creates two or more fully intertwined systems.

This, then, is the coevolutionary argument about complex adaptive systems in a nutshell. Because of his expertise, Norgaard initially focused on the economic side of social systems as it allowed him to operationalize and measure coevolution. He incorporates the ability of decision-makers to deliberately select and manage the selection pressures that feedback loops exert on systems and to decide on the pool of resources, i.e. how they are utilized. Thus, decision-makers do not just respond passively to pressure, and decision-making in coevolution means dealing with feedback patterns in terms of selection and selection pressures (cf. Foster, 2005; Foster & Hölzl, 2004).

There is no reason why the application of coevolution should remain restricted to socio-ecological systems. It is possible to map feedback loops between any policy decision and the resulting changes in its environment, and examine how different actors adapt to changing circumstances (Sotarauta & Srinivas, 2006; Manner & Gowdy, 2010; Waring, 2010). Coevolution is a versatile concept that can be applied in multiple ways, ranging from Norgaard's analysis of intertwined social and ecological systems to institutional and organizational change (Hodgson, 2010) and genetic-cultural evolution (Hird, 2010). Among others, Ayres (2004) argues that each application requires a specific operationalization of the different mechanisms of coevolution, in particular reciprocal variation, selection and retention, to understand how systems change and in which direction they change. This does not mean copying the notions from evolutionary theories without considering the differences between, for example, biological coevolution and public decision-making (ibid.). Instead, we need to build a specific conceptualization for our quest by combining the different strands.

5.2.2 COEVOLUTIONARY MECHANISMS

The explanation above indicates that coevolution involves reciprocal selection. This begs the questions: what is selected and how is it selected? We will answer both questions in this section. Following John (1999), we understand that public decision-makers have certain ideas they want to follow up and with which they compete with each other. These ideas concern both problem and solution definitions. Problem definitions refer to the way that issues are perceived and framed so that they become policy problems, while solution definitions refer to the way solutions to these problems are envisaged.

The attempts to regenerate Nieuw-Crooswijk, an old district in Rotterdam, are a useful illustration (cf. Edwards & Schaap, 2006; Eshuis & Edelenbos, 2009). The district did not have a good reputation after the Second World War and comprised mainly obsolete and cheap rental houses. The city council attributed the social problems in the district to the skewed allocation of social housing. Thus, it developed a plan to buy out the families in those houses, tear many of them down, and replace them with middle-class houses and green spaces. However, some of the inhabitants argued that, while they shared the council's concerns about the social issues facing the district, they did not believe that large-scale demolition was the answer. Other residents argued that neither the problem definition nor the solution definition were correct. This example shows that actors may share certain problem and solution definitions while disagreeing on others. Following Kingdon (1986), solution definitions may precede and perhaps even determine problem definitions.

For the sake of convenience, we use the abbreviation 'PSC' to denote the specific problem-and-solution combination as used by each actor. The PSC is our unit of selection. Actors make decisions that have a real impact on the world based on the prevailing PSC. In the case of Nieuw-Crooswijk, the city council's PSC attracted enough support to lead to a decision in favor of large-scale demolition and the construction of more expensive houses. This created a number of effects, such as the closure of some shops and the departure of families from the district, while others moved in. Some of these changes were intended, while others were not. The financial crisis of 2008 resulted in a drying up of credit available from the banks. This affected the main property developer in the area, as well as families seeking a mortgage to buy a house. Ultimately, the crisis led to a severely scaling down of the original building plans. Since a considerable part of the area had already been torn down, this change left large vacant patches of land between the old and new buildings.

Cafe Bar
My World
CITYTEC
272 00 00

This current situation can be explained through the systemic nature of feedback and chance developed in the previous chapters. Essentially, a new situation has emerged to which decision-makers have to respond: they will have to reconsider their PSC in the light of that new situation. Feedback loops have been created between the prevailing PSC and subsequent policy decisions, tangible responses to those decisions, the occurrence of chance events, and the formation of new PSCs. We can now answer the questions as to what was being selected and how it was selected.

Variation and selection occurred at two levels: the level of the PSCs and the level of the system states. At the PSC level, variation and selection occurred as different PSCs competed with prevailing PSCs over times. This variation was affected by real world events and the dynamic responses to the decisions made. In the case of Nieuw-Crooswijk, for example, we find that the economic crisis hampered the plans, presenting the decision-makers with a situation that rendered their PSC obsolete. Variation and selection also occurred at the level of the system, since certain decisions rendered certain possible future states more possible or, conversely, out of reach. The decision to buy out families and demolish many old houses has made it possible to build new houses, but has also made it impossible to return the district to the old situation or state, even in the hypothetical case where it is found that the wrong decisions were made. The creation and selection of variety is therefore considered to be part of the same loop: variation and selection are intertwined (Foster & Hölzl, 2004). PSCs and consequent decisions generate variation and selection in the real world; the real world generates variation and selection in the PSC's and consequent decisions.

Over time, the degree of freedom to act is shaped by the pattern of variation and selection. Making decisions based on the PSC means exerting selection pressures on the system. Responses from the system in turn mean exerting pressures on the PSC and decision-making. Norgaard (1994) equates selection pressures with feedback because they determine future possibilities in the system. Such feedback loops carry information for actors, which is assessed and acted upon, as discussed in the previous chapter (see e.g. Dopfer, 2005; Foster & Hölzl, 2004 in the context of evolutionary economics). Which particular PSC comes out on top of the competition is therefore not a given (John, 1999). It depends on the feedback loops and their cumulative effects, as well as the strategies followed by actors, e.g. the support they manage to amass by cooperating or going for a stand-alone strategy if they think this will bring them closer to their goals.

It is through this coevolutionary focus that our understanding of complex adaptive systems, feedback processes and human agency comes together. While the core of coevolution, namely reciprocal selection, concerns content (i.e. what is selected and how that impacts the state of the system), the coevolutionary approach adopted by Norgaard, among others, introduces elements of structure (complex adaptive systems and agency) and process (positive and negative feedback following variation and selection, punctuated change, hysteresis, path-dependency and lock-in). This wide variety of concepts needs to be logically integrated into a coherent framework, and we utilise the idea of fitness landscapes to do so.

5.3.1 THE ALLURE OF THE FITNESS LANDSCAPE

A powerful means of summarizing and visualizing the argument above is by developing a so-called fitness landscape. The idea of the fitness landscape can be traced back to Wright's work on biological evolution (1932). He assigned a value to each genotype to denote the match or fitness between that genotype and its environment. The distribution of these values across the space of genotypes constitutes a fitness landscape (Kauffman, 1993). A fitness landscape is often visualized as a three-dimensional rugged field or landscape, or as a hypercube. This is helpful in showing that the change in each so-called agent in the landscape is driven reciprocally through its interaction with other changing agents, which we now know is a form of coevolution (Kauffman & Johnsen, 1991). Fitness landscapes show how the position of an agent in the landscape is conditional because this position is derived from the relative distance between other agents. As Arthur & Durlauf put it, "*how* individual agents decide what to do may not matter very much. What happens as a result of their actions may depend much more on the interaction structure through which they act—who interacts with whom, according to which rules." (1997: 10; italics original). While in general we do not agree that how individual agents decide is not important, notice to the discussion in the previous chapter, we do agree that we need to find a way to understand the volatility of decision-making in a system and a fitness landscape is a helpful heuristic device for that purpose (cf. Plutynski, 2008).

If we assume that agents in a system are constantly searching for a better 'fit' with their environment through a space of possible combinations and outcomes, the notion of reciprocity implies that the environment will move simultaneously with the search for that improved fit. Search takes place in a dynamic environment where peaks in the (rugged) fitness landscape constitute an improved 'fit' or local optimum. In addition, the more autonomous the agents, the more rugged the landscape is. The dynamics of the landscape imply that the peaks emerge out of re-combinations of agents, and what constitutes an optimum at any given point in time may turn into a valley later on because of the dynamics in the landscape. Local search processes towards improved fit can be likened to hill-climbing, but with the understanding that any hill or peak is temporal. In fact, in the search for optimization "[...] many parts and processes must become coordinated to achieve some measure of overall success but conflicting "design constraints" limit the results achieved." (Kauffman, 1993: 33)

Kauffman calls the fitness landscape a transparent metaphor for understanding coevolution in biological systems, but not an accurate explanation of reality. We need to remind ourselves the lessons of theory transfer from Chapter 1: something that works in the source domain may not work in the target domain. For us, the fitness landscape is primarily a heuristic device. Different adaptations of the fitness landscape have been used in non-biological domains. The example from Frenken and Nuvolari (2004) discussed in Chapter 2 shows how it can be used in an evolution-based analysis of economic systems (see also Arthur & Durlauf, 1997; Frenken, 2006). Other applications include ecosystems (e.g. Kauffman & Levin, 1986; Levin, 1978; Fath & Grant, 2007) public policy (e.g. Room, 2011) and game theory (e.g. Lindgren, 1997), from which Kauffman also borrows.

A particular adaptation of the fitness landscape that is more appropriate for social systems and situations is the so-called performance landscape (Siggelkow & Levinthal, 2003), which maps the attempts of competing companies to maintain internal coherence while simultaneously trying to achieve a better fit with their business environment by "identifying activity configurations that are not only internally reinforcing but also appropriate, given the firm's current environment. For incumbent firms, this challenge is particularly acute after environmental changes [...] that allow *new* ways of performing *existing* activities (2003: 650, italics LG). Rhodes (2008; Rhodes & Donnelly-Cox, 2008) adapted the performance landscape to demonstrate the interrelatedness of strategic choices by actors and outcomes in the context of the performance of housing policies in Ireland. The model was developed in consultation with stakeholders and showed that the "[...]relationship between strategic choices and performance outcomes could vary over time and, in some cases, could be influenced by the efforts of one or more agents within the system [...] (2008: 364). Viewing public decision-making through the lens of a fitness or performance landscape clarifies that: [a] an agent's ability to achieve its goals depends on its position relative to other agents, not just its intentions or deliberate design, and [b] the process of coevolution as described in the previous section determines the outcomes. However, this metaphor should not be directly applied to social systems without any modification because the properties of such systems are harder to determine indisputably compared to biological systems (Gill, 2008). Questions that arise when the metaphor is applied to social systems include: what exactly constitutes an improved 'fit' when social actors are involved? How can an actor's goals be unambiguously established? Since fitness landscapes originally relied on quantitative values to indicate the varying levels of fitness of interacting genotypes, how can the relative distance between actors be similarly quantified?

The above issues need to be taken into account when a fitness landscape is being applied in our context. Our fitness landscape must meet a number of specific requirements. First, it must be able to show how the interactions (or the lack of them) between actors operating in a particular policy issue reciprocally influence other actors, i.e. how do these interactions alter the space of possibilities. Second, it must account for the content of public decision-making, which we defined earlier in terms of problem-and-solution combinations. Third, it must incorporate the feedback effects resulting from decisions and the occurrence of chance events. Fourth, the previous points mean that the fitness landscape must be able to show changes as time progresses. Fifth, it must feature measurable properties. Altogether, it must account for the identifiers of complexity mentioned in the introduction of this chapter: structure, process and actors, and temporality, adaptation and diversity at the intersections of structure, process and actors. The following section describes the specific representation that results from these requirements.

5.3.2 A FITNESS LANDSCAPE FOR PUBLIC DECISION-MAKING

The landscape we model should not pertain to public decision-making in general, but should instead be context-specific. Thus, we restrict it to the particular issue about which decisions are made, such as 'urban regeneration in Nieuw Crooswijk', as per the case introduced above. We will want the label to be broad enough to encompass all possible perceptions of this issue, yet not be so generic that it could include everything other than urban regeneration in this particular city district. A convenient method to achieve this is to base it on the maximum variety of perceptions present in a particular case. This sets the boundaries for the fitness landscape. We will later look at what happens to those boundaries if a new actor with a different perception enters the landscape. Our fitness landscape of urban regeneration in Nieuw-Crooswijk will be three-dimensional, i.e. it will have three axes: x, y and z.

Fitness landscapes, in the original meaning of the word, were populated by agents. 'Agent' is a convenient term for any information-processing entity, such as a gene. In the interests of consistency with the rest of this book, we replace the word 'agent' with 'actor'. In fact, 'actor' is an ambiguous word in the field of public administration and public policy. It can refer to entities as diverse as individuals, teams in organizations, and inter-organizational groups (cf. Koppenjan & Klijn, 2004). We want that diversity to be part of our fitness landscape. From the discussion in Chapter 4, we understand actors to be groups of decision-makers that self-organize according to the principles of autopoiesis. They could be organizational units, but not necessarily so, as argued in that chapter. Actors are treated as singular units on the condition that the constituent individuals consider themselves to be a coherent whole. This coherency arises out of their communications, expectations and a shared sense of belonging. It allows us to treat actors from different constituencies similarly in our fitness landscape.

We will return to the nature and role of actors later on. The next step is to position the problem-and-solution combinations, or PSCs, on the *x*-axis. As discussed in Section 5.2.2, problems and solutions are often coupled in practice. Thus, we consider them together in our model in the specific configurations as proposed and advocated by the different actors in the landscape. In actual policy debates, various combinations and problem and solution definitions are possible, and to differentiate each particular PSC as advocated by certain actors in quantitative terms would be impossible without the use of elaborate methodsological operations. We opt for a more intuitive approach. It is possible to differentiate between PSCs that are of a more singular nature and PSCs that are of a more composite nature. Singular PSCs lie at one end of the *x*-axis and their problem and solution definitions have a very limited scope. In the case Nieuw-Crooswijk, for example, the problem could be defined as degeneration being caused by the presence of cheap houses, and the solution defined as demolition. Composite PSCs lie on the other end of the *x*-axis and their definitions account for more variables. For example, actors may define the problem of urban degeneration in Nieuw-Crooswijk as the result of a mix of related factors and, consequently, offer a set of related possible solutions. Other combinations, such as a composite problem definition combined with a singular solution, can be found between the two extremes of the *x*-axis.

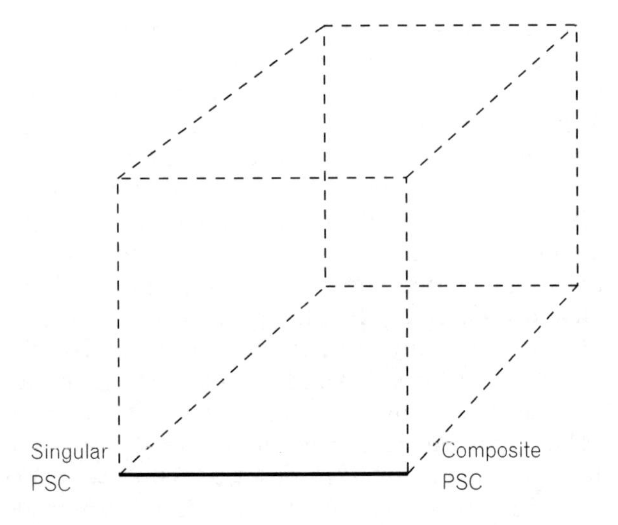

Obviously, the singular or composite nature of the PSCs is not the only determinant factor in public decision-making. Two different actors may think in terms of a singular PSC but still feel uncomfortable with each other, for example if they have different solutions. In the Nieuw-Crooswijk case, one singular solution could be the demolition of all of the houses, while the other solution could be to preserve them. Although singular, the two solutions are very different. Thus, the actors experience or perceive a distance between themselves and others, regardless of the distance between them as mapped on the *x*-axis. To account for these differences in perceptions, the *y*-axis represents each actor's perceived degree of belonging and internal coherency. Assigning a position on the *y*-axis to an individual actor requires knowledge of how that particular actor rates the degree of belonging with other actors in the field. When measuring this, one needs to account for the different perceptions that different actors may have of their distance from each other. This is resolved by deriving the distance as an average of both of their perceptions.

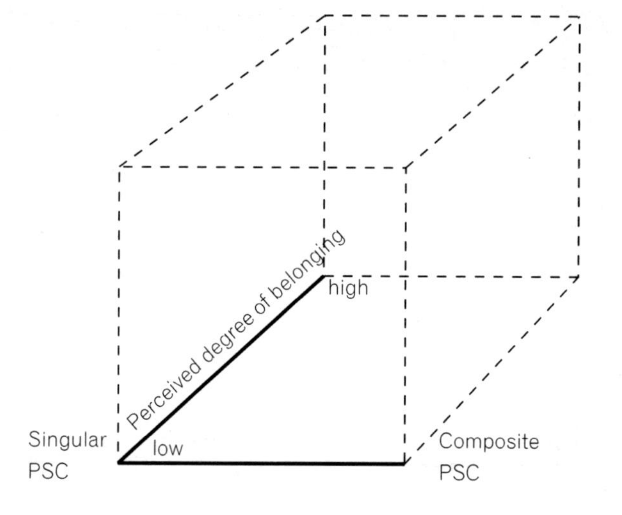

We have now created a two-dimensional landscape that allows us to group different actors relative to each other. This takes into account the content of decision-making and the autopoietic nature of actors in establishing and maintaining themselves. Moreover, we can treat actors of very different sizes in the same way. Over time, actors will converge and diverge in their attempts to gain a better position or fitness. The core of the fitness landscape lies in the way it relates one's choices and adaptive moves to the degree of fit with the environment, i.e. with the system characteristics. The extent to which actors succeed in aligning their PSC with these characteristics is expressed with a higher position on *z*-axis, i.e. a higher level of fitness. A higher position indicates that an actor is closing to achieving its goals, i.e. the realization of its PSC, and a lower position means that an actor is further away from achieving its goals.

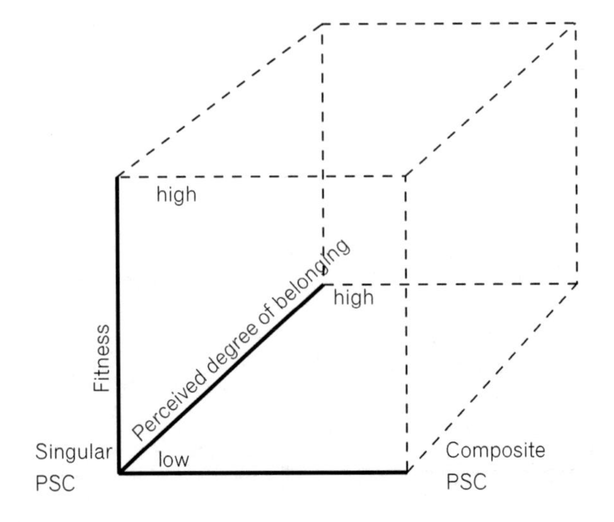

Opting for alignment as a measure of fitness rather than goal-attainment (cf. Venkatraman, 1989, for other variants) helps us account for the fact that actors' preferences and possibilities coevolve with the system's dynamics, as argued in Section 5.2.2. For example, an actor's PSC may have been rendered obsolete because of some unforeseen events in the system, in which case the achievement of a PSC is impossible or irrelevant. Note that a PSC does not need to be centered on the core issue on which the fitness landscape is build. In our example of Nieuw-Crooswijk, an actor's PSC may include a search for improving one's legal status, and urban regeneration may be seen as a means to that.

Earlier, we stated that the landscape is defined by the maximum variety of perceptions present in a particular case. It is conceivable that new actors may enter the landscape or that some actors may leave the landscape. We have to make a distinction between three possibilities. The first possibility is that the new actor fits within the given limits of the landscape, in which case nothing much changes in the dimensions of the landscape, even though the new actor may lead to different interactions. The second possibility is the arrival of a new actor with such a different PSC that it does not fit within the existing landscape. If other actors respond to this new actor by changing their PSCs and adaptive moves, the landscape will also change accordingly. If, however, no actor responds to the new entrant in any way, the new actor is in fact not an actor in this particular landscape. The third possibility is that an outside actor does something that influences the PSCs of the actors within the fitness landscape without becoming involved in this particular issue. For example, if the national government demands that the Rotterdam city council builds more houses in the city, and the council decides to do this in Nieuw-Crooswijk, the national government does not become a new actor. Instead, it changes some of the conditions of the landscape and the PSC of the council, if the council decides to build in Nieuw-Crooswijk. Thus, the nature of the new actor and the responses of the existing actors determine how the fitness landscape alters because of the new entrant.

We have now created a three-dimensional fitness landscape which we can use to map the actions of different actors in the complex world we described in the previous chapters. In particular, it allows us to see how the position of actors is influenced by the adaptive moves of others and the complex dynamics that take place in the system, and how this in turn coerces actors to engage in further adaptive moves. What may appear to be a clear route towards higher fitness may turn into a dead end later because the environment has evolved in a different direction. The next sections show how the dynamics of complexity can be mapped in the fitness landscape.

5.3.3 DYNAMICS IN FITNESS LANDSCAPES (1): ADAPTATION, DIVERSITY AND TEMPORALITY

At the beginning of this chapter, we identified how the coming together of structure and process is characterized by temporality, the coming together of process and actors is characterized by adaptation, and the coming together of actors and structure is characterized by diversity. These identifiers, though they are too coarse for summarizing a nuanced narrative, help us to structure the many aspects of dynamism that can be derived from fitness landscapes. Following our discussion in Chapter 1, it should be stressed again that there is no natural or given starting point in such an expose. However, as public decision-making is ultimately a human act, even if it is a severely constrained one, it does make sense to start with the role and operations of actors, i.e. *adaptation*.

Actors are in a constant search for alignment between their environment and their own PSCs. As the environment evolves because of the cumulative impact of the feedback loops in operation and the adaptive moves of other actors, the actors will need to adjust themselves to make alignment more likely. However, the situation could change to such an extent that alignment suddenly becomes more likely than before. In short, the adaptive moves of actors and the feedback loops in the system influence the current position of actors and their space of future possibilities. The adaptive moves can be seen as attempts to reach alignment under changing conditions. As argued in Chapter 4 and earlier on in this chapter, the adaptive moves include a search process to find opportunities for alignment, and decisions that are supposed to bring the actor closer to alignment. While the individual processing of information cannot be mapped in the fitness landscape, we can observe the resulting changes, particularly the changes in the PSCs and the grouping or ungrouping of actors if they perceive that doing so helps them achieve alignment.

Here, we return to the concept of policy action system, as presented in Chapter 2, to understand what adaptive moves entail. Actors will group themselves in policy action systems as they converge and diverge over time. For example, they may think it is useful to group themselves with other actors if those actors share a similar PSC, or if another actor possesses resources that can help achieve alignment. An actor leaves a certain grouping if it thinks that the other actors are no longer helpful in achieving alignment, or if its own PSC has changed to such an extent that it is no longer relevant to stay grouped with them. There could also be instances where an actor with a changed PSC may try to persuade others to follow a new course. Over time, a policy action system can become more homogeneous if the actors in the system manage to mutually align their PSCs, leading simultaneously to greater internal coherence and external closure. Conversely, a policy action system can become more heterogeneous if it consists of actors with different PSCs who deem it necessary to cooperate despite their differences. This means that internal coherence is loose, perhaps even fragile, but is countered with more external openness. One may argue that a heterogeneous policy action system has a higher chance of detecting and adapting to changes in the environment. Yet, at the same time, homogeneity allows actors to focus on aligning the PSC with the environment. How homogeneity and heterogeneity are related to alignment is ultimately an empirical issue.

The argument above means that adaption and the generation of *diversity* are intimately linked. Adaptive moves of actors into and out or policy action systems and the nature of those action systems generate and limit diversity in the various PSCs and future possibilities. The generation of diversity influences adaptive moves simultaneously as it may open new avenues of possibilities, whilst at the same time closing down other possibilities, i.e. the argument of reciprocal selection as discussed in Section 5.2.2. Because of these dynamics, each current situation holds a space of possibilities (Mackenzie, 2005): a number of possible future system states, some more likely than others. There is a certain bandwidth of plausibility as to what may happen next. For example, while it is more likely that Nieuw-Crooswijk will become a middle-class neighbourhood than an office park, the latter is not impossible if that would help the property developer when it becomes increasingly hard to sell middle-class houses. The plausibility rests in the dynamics of the world: which events enable certain futures whilst closing down other futures, and which adaptive moves mean that some goals become attainable and others less so.

Coming full circle, the arguments of adaptation and diversity, as expressed in the limitations and possibilities of future system states above brings us to the third aspect, namely *temporality* at the intersection of processes and structures. In fact, the narrative of complex systems cannot be told without the time dimension. Dynamism rests on the fact that time progresses and that each situation is slightly or significantly different from the previous situation. Feedback loops need time to become actualized, responses take time to gather and become effectuated. Feedback delays play an important role in the process of adaptation and the generation of diversity because actors need to rethink their position and possibilities. As time progresses, actors and their environment coevolve accordingly in continuous attempts to attain fitness.

5.3.4

DYNAMICS IN FITNESS LANDSCAPES (2): RUGGED LANDSCAPES AND ADAPTIVE WALKS

If we map actors and their attributes as discussed in section 5.3.2 at a three-dimensional fitness landscape, we end up with a field resembling a mountainous landscape because each position on the *z*-axis can be considered a peak. As we have seen above, the interdependence among actors in the landscape means that the adaptive moves of one have repercussions for the other. Naturally, if one actor or group of actors manages to get closer to the alignment of their PSC's with the environment, the actors supporting different PSCs become more distant from alignment. A closer alignment of one actor is inversely related to alignment of other actors. In our three-dimensional landscape, this means that if an actor achieves a higher position on the *z*-axis, other opposing actors remain at their current position or may even obtain a lower position. What is a peak to one actor represents a valley to others.

In Kauffman's conceptualization (1993), the level of fitness is influenced by the degree of interdependence between actors: the fewer the interactions, the fewer the peaks in the landscape (Levinthal & Warglien, 1999). This suggests that higher levels of fitness can only be obtained if the agents interact. Kauffman derives this mechanism from game theory and builds on the assumption that, ultimately there is one optimum for the fitness landscape as a whole. Consequently, he can calculate Nash-equilibriums for individual agents within that given optimum. That does not fully work for the type of cases in public administration and public policy for two reasons. First, 'interactions' can mean many things in our field. It could be full cooperation, but also the exchange of resources. Actors who meet in court are interacting, as are actors who are engaged in a stakeholder dialogue; but both types of interaction are arguably quite different. Second, the issues we consider can have multiple local optima, depending on the PSCs held by the individual actors. Actors who aim to carry out a policy plan have a different optimal outcome than actors who aim to obstruct the same plan.

Consequently, we retain Kauffman's idea that interactions matter in achieving fitness, but do not hold that a higher number of interactions automatically equals a higher level of fitness. Moreover, we need to account for multiple optima within the given fitness landscape, as actors will cluster in policy action systems if they think that clustering can contribute to a higher individual level of fitness. This could mean that actors with different PSCs could cooperate because they believe that they can both gain from cooperation. It could also mean that an actor may opt to go alone if it thinks that its current interactions hinder it in achieving fitness. Over time, the fitness landscape will be one of evolving peaks and valleys. If all actors present in the field agree on a single PSC and have shared perceptions of mutual distance, the landscape would consist of a single peak, i.e. one single optimum. If, in the more likely case, multiple actors have multiple optima, the landscape becomes rugged with multiple and changing peaks (Levinthal, 1997).

The clustering and de-clustering of actors, as described here, constitutes our version of the adaptive walk (cf. Kauffman & Levin, 1987; Levinthal & Warglien, 1999). An adaptive walk summarizes the route an actor takes through the fitness landscape, whereby it interacts with fitter actors to improve its own position, and ultimately, to get ahead of its competitors. In principle, this means that actors have to search for the best route towards their own peak in the landscape. In their adaptive walks, actors form clusters around peaks. But, as described above, the mutual dependencies of actors' positions means that what constitutes a peak at one point in time may become a valley at the next point in time, because when (clusters of) actors make a move that improves their fitness, they may diminish the fitness of others at the same time.

As a consequence of these coevolving dynamics, individual actors must carefully map the pathways towards improved fitness and recognize that their intended pathways may be substantially altered during their walk as other actors respond to that walk. This implies a continuous search process across the space of possibilities to seek out opportunities for improved fit. But as we have argued before, actors can only deploy a limited, myopic search process. Consequently, they will only scan the nearest opportunities and what may appear to be an opportunity nearby may in fact be a dead-end road in the long run because other actors have moved in a different direction and the environment may have evolved in a different way.

The theory of adaptive walks also takes into account the gradient of the walk: the steeper the inclination, the more difficult it becomes for actors to reach the peaks. This is a tempting idea that gives us a first approximation as to how the level of difficulty of different pathways can be expressed and compared. However, the model we have presented is not accurate enough to calculate gradients of routes and thus does not allow to make somewhat precise estimations of such pathways. What adaptive walks do tell us is that what may appear to be a straight route to better fitness may in fact result in the actor arriving at a position from which it cannot move any further towards fitness without first 'climbing down' the slope.

5.3.5 DYNAMICS IN FITNESS LANDSCAPES (3): ATTRACTORS AND TRAJECTORIES

The discussion in the previous sections has constantly alluded to the fact that actors coevolve with other actors and with their environment. As a consequence of these mutual adaptations, systems evolve over time. An element of each fitness landscape is the system's state at any given time. A system state can be considered an attractor in an attractor basin comprising all possible future states of a system (e.g. Arthur & Durlauf, 1997; Byrne, 1998; Levinthal & Warglien, 1999; Mackenzie, 2005; Martin & Sunley, 2006). It helps describe the current state of the system and map the bandwidth of possible and plausible futures. As not every system state is very likely in the future, "[...] attractors "box" (limit – LG) the behaviour of a system into small parts of its state space or space of possibilities" (Kauffman, 1993: 174). As such, each system state at a given point in time has more and less likely futures.

The word 'attractor' has the connotation that it attracts the system to a certain state. However, an attractor is not a 'thing' out there, but an expression of the extent to which a system can change, indicated by its resilience against disturbances (Marion, 1999). Hypothetically, a system may always return to exactly the same state when under pressure, expressed by a fixed point attractor (Otter, 2000). A periodic or torus attractor describes the alternation of systems between a limited number of states (Mackenzie, 2005). While this may be a more realistic description of social systems, our earlier discussion of hysteresis (Chapter 3) and time-asymmetry (Chapter 4) highlighted the inability of a social system to return to any of its previous states – each new state is ever so slightly different from the previous one. This situation is expressed by the strange attractor (see also Byrne, 1998), which represents the temporally stable state of equilibrium. It can remain in this current state because of existing feedback loops that maintain a particular situation. It may take considerable pressure before the system moves into another stable state. Thus, the changes from one attractor to another may happen in a punctuated fashion. Exactly which next strange attractor in the attractor basin becomes actualized depends on the local and temporal circumstances: the feedback loops present in the system, the adaptive moves of actors and their PSCs as described above. The next system state is therefore configured by the conditions present in the current system state.

The sequence of system states describes a trajectory as a result of the coevolutionary processes. What can be said about the direction of those trajectories? Do they move in a certain direction or end-state? Does a change equal improvement or aggravation? According to Kerr (2002), confusion over this directional dimension has resulted in diverging interpretations of the nature of coevolution in social systems. Arguably, it is important for actors in a fitness landscape to sense the direction in which the system is evolving. A survey of the literature suggests that the directionality of coevolution is explained in four diverging, but sometimes converging, views on coevolution: coevolution as progression, coevolution as equal distribution, coevolution as the result of intentional action, and multidirectional coevolution.

The directional demand concern a succession of states, but not necessarily a progression to a better or more favourable state. However, such interpretations exist and coevolution from this perspective is regarded as a non-linear route towards increasing improvement (Kerr, 2002). These interpretations are, perhaps, informed by an interpretation of the Darwinian thesis of the survival of the fittest, i.e. that evolution means that an actor will either become better or lose relevance. However, both Kerr (2002) and Sanderson (1990) argue that there is no fixed relationship between evolutionary change and progress as improvement. What constitutes an improved state is based on subjective judgements and what is seen as progression by one actor may be regarded as regression by another. As the environment continues to evolve, the supposedly better fit may be lost because the conditions of that fit have changed, as shown in the argument about alignment between PSCs and the environment. Norgaard traces such perceptions on improvement back to a materialist ontology. He argues that this perspective actually obstructs a thorough understanding of the processes of coevolution (Norgaard, 1994; 1995).

A further development of coevolution as improvement concerns accounts in which coevolution is regarded as an equal distribution of the burden between different actors or between different systems. The interaction between different systems that leads to coevolution is therefore seen as a desirably balanced one between the two (cf. Ruijgrok, 2000). This implies a normative approach to coevolution, i.e. that developments in one system should not exploit resources in another system if it is impossible to replenish these resources. Such a goal is praiseworthy in itself but should not necessarily be labelled as coevolution. Proposing non-interference as a requirement in the connected development of systems rules out the possibility of interference during coevolution, even though such interference is probably caused by reciprocal selection, as argued previously. Since reciprocal selection lies at the heart of coevolution, it is contradictory to this normative explanation.

A number of accounts revolve around the idea that coevolution is something that actors can create by lifting the interference. The concept of conditional coevolution, i.e. coevolution that can only come into existence when actors develop the right circumstances, appears to have gained ground in accounts of complexity from the field of management science. Two recurring themes in this perspective are mutual influence and cooperation for the benefit of all concerned. Conditional coevolution is rooted in the idea that hierarchical relations between actors should be replaced by relationships that are more network-like. Cooperation between actors should then help to establish coevolution, which means that the participants engage in a mutually favourable interaction. Once again coevolution is given an exclusively positive connotation. However, as discussed before, reciprocal selections take place regardless of the intentions of actors and regardless of whether the outcome is favourable to all concerned. Hence, coevolution is not something positive that exists only because of management incentives.

Clearly, the argument in this book builds on coevolution as a neutral, explanatory theory that does not say much about its direction or even desirability. Consider, for example, the research into the lasting instability in coordination over urban planning in the Amsterdam metropolitan region by Schipper & Gerrits (2012). This longitudinal study analysed the different system states of coordination between 1982 and 2007 as a result of the coevolution between the system and the environment, and the way that the actors dealt with the selection pressures they were subject to. The long time-span of 25 years allowed the researchers to conclude that instability was not an experience at one given point in time but rather a constant in the way metropolitan coordination responded to changes in the environment in attempts to gain and regain fitness. It was subjected to chance events, such as a mandate by the national government to accommodate 150.000 new houses, and the current form of coordination in the shape of the Metropoolregio Amsterdam (MRA) was not designed but emerged out of a series of conferences that were that were the unforeseen outcomes of previous adaptations. Along the way, coordination took on many different appearances, ranging from non-committal meetings to the assumption of legal powers and budgets. The study demonstrated that coevolution (in the form of the succession of system states) could take place without a particular end-state being predefined. Each particular step may be an improvement for certain actors and present a worsening to others. The value of coevolution is that it offers a reciprocal perspective on change, or lack thereof, in the systems that are governed.

5.4 CONCLUSIONS: A COEVOLUTIONARY REVISION OF PUBLIC DECISION-MAKING PROCESSES

The core theme of this chapter is to characterize the dynamics of public decision-making in terms of coevolution. We used the concept of the fitness landscape to illustrate coevolutionary dynamics and its consequences for actors, whilst acknowledging that actors themselves play an important role in those dynamics. It should be emphasized that the fitness landscape is first and foremost a heuristic device that is helpful in illustrating and discussing the various dynamics that surround public decision-making in complex systems. As such, it helps outline the local optima, the different pathways to optima through grouping and un-grouping, and the mechanisms that render an optimum easier or more difficult to reach. A fitness landscape does not hold explanatory power, but is more of a vehicle to demonstrate the different mechanisms and concepts discussed in the previous chapter. While its strength lies in its flexibility to accommodate these mechanisms and concepts, its potential weakness is that it can be under-specified. The reader is invited to play around with it and to tinker with its dimensions to see what happens as a result. The fitness landscape as modeled in the previous sections gives plenty of room for that. As a heuristic device, it guides intellectual exercises about the dynamics of public decision-making.

Five chapters ago, we began the argument by stating that complexity is not a truism but a very real property of public decision-making. Having gone through so many ideas, theories, mechanisms and cases, where does this leave us? Coevolution shows that decision-making processes are subject to patterns of reciprocal selection and that decision-makers can be both architects of new stable and favorable states as well as victims of their own decisions. The apparent rational act of setting goals and routes towards those goals is hampered by the dynamics of the real world. It may mean that attempts to reach those goals are futile, but also that certain goals may become attainable sooner and more easily than expected.

There are five aspects of public decision-making in coevolving systems. First, it appears that although public decision-makers expect that their choices and actions to lead them to their desired future attractors, they are also subject to chance events and feedback effects in the system that can disrupt good intentions and careful planning. These feedback effects include responses to earlier decisions. Thus, the attractor basin is not only limited or expanded through intentional actions, but above all because of those effects. Decision-makers respond to the actual situation rather than remaining unrestrained and being able to steer as they desire.

Second, there appears to be an erratic relationship between the decisions made by the actors in the policy action system and the subsequent responses in the system it attempts to govern. These responses do not evolve gradually and regularly with the actions from the policy action system but instead show a punctuated nature, with changes taking place elsewhere in place and later in time, and with erratic results. Therefore, the policy action system could face a new situation relatively unexpectedly. The relationship between the decisions and the actual outcomes is obscured because of the complexity of the feedback patterns that surge through the system. Together, this often renders change unintended and sometimes unexpected.

Third, when facing uncertainty as presented by a world that does not willingly comply with their intentions, decision-makers deploy various decision rules and heuristics and organize in policy action systems that are more homogeneous or heterogeneous. Homogeneous systems favor connections with those actors who support the existing problem-and-solution combination, and aim to shield the decision-making from those who oppose it, in an attempt to keep the process under control as it is considered complex enough as it is without any additional distractions. This results in a narrow PSC. Heterogeneity emerges when a homogeneous response is no longer feasible or deemed tolerable. Heterogeneous systems are more open to connections with actors with different PSCs to expand the diversity of ideas and goals in the decision-making. This results in debates around the current PSCs.

Fourth, while the classification into homogeneous and heterogeneous policy action systems may suggest a stable dichotomy, it has been observed empirically that heterogeneity is enclosed within homogeneous policy action systems but is not always unlocked. By definition, policy action systems are made up of multiple actors which means that the potential for homogeneity or heterogeneity is always present. Policy action systems appear to alternate between the two extremes over time. The change or consolidation of a regime is induced by actual unfavorable events or by the perceived imminent risk of such changes. While a change or consolidation of a given state may be a deliberate response to current pressures (that could stem from earlier decisions), it has also been observed that both types of responses have the capacity to reinforce their nature unintentionally. The homogeneous policy action system is driven by its self-referential nature that reconfirms its workings while the heterogeneous policy system is driven by further dissipation in an attempt to be comprehensive. The nature of the policy action system can change partly uncontrollably, as a homogeneous policy system may not be aware of its singularity and a heterogeneous policy system may not be able to keep its diversification under control.

Fifth, when all of these points are combined, it appears that the feedback loops or selection pressures of coevolving systems have a reciprocal quality insofar as the degree of freedom of the policy action system is limited by events and developments outside the direct, perceivable and intended control of the actors within the system. The attractor basin containing the possible future states of the systems could be compromised through adverse, unintended results and events, while the policy action systems could also be only partly in control of their regime in handling the ensuing pressures.

These five aspects summarize what the coevolutionary revision of public decision-making means. Coevolution between systems is indeed a matter of reciprocal selection, with the results not being fully determined by the intended selections made by decision-makers, but rather emerging from the complex process of reciprocal selection. This is a reminder of our limits in shaping the world. This is not the place to become cynical or desperate about what public decision-making can achieve. Indeed, much has been achieved. Complexity is not about giving up but about accepting that circumstances, situations, desires, goals and outcomes always evolve over time, as demonstrated by the cases discussed in this book. Complexity is not external to the decision-maker and cannot be switched on and off at will. Accepting this means that a decision-maker must equally be able to shape reality as to adapt to new circumstances, and to accept that some futures will not happen or that certain policies may have missed their mark. Indeed, even the controlled collapse of a certain policy or project may have its merits. The most cynical response to complexity is therefore to ignore it, to pretend that it does not exist. This book hopes to have convinced readers that complexity can and should be known and understood. Ultimately, this understanding will lead to better decision-making and better policies. Exactly how complexity can be researched and how system states and system trajectories can be researched is subject of the next chapter.

CHAPTER 6.

BEING INQUISITIVE: RESEARCHING COMPLEXITY.

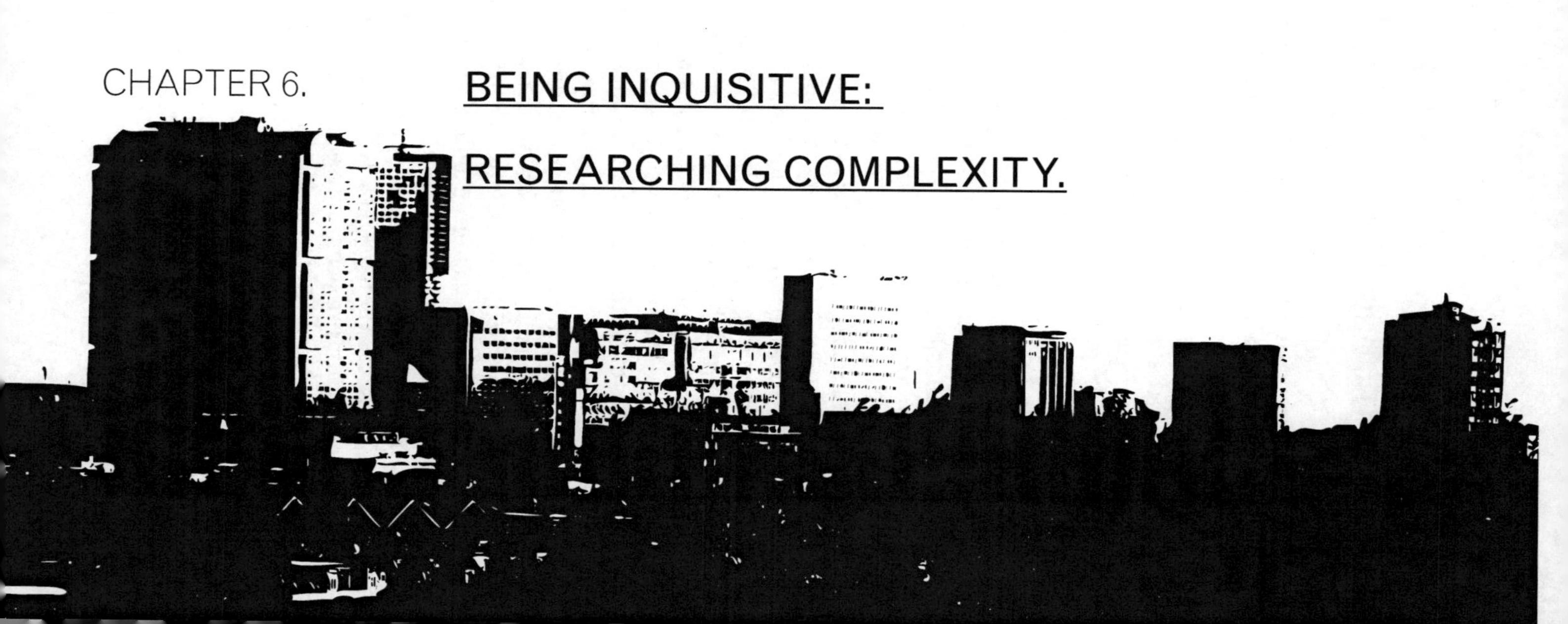

6.1 INTRODUCTION

It is imperative for any scientific account to be explicit about the foundations on which its claims are made, and to keep the claims coherent and consistent. The previous chapters presented a specific view of the complexity of public decision-making that is rooted in a realist worldview. Nonetheless, researchers have probed the complexity of public decision-making from various angles, using diverse methodologies such as qualitative, interpretive analysis (e.g. Wagenaar, 1997; Wagenaar & Cook, 2003) and computational modeling (e.g. Koliba & Zia, 2012; Koliba, Zia & Lee, 2011). The common goal of these efforts has been to uncover that complexity. Putting it metaphorically, we researchers are stumbling about in different corners of the same murky basement, looking for the light switch that will illuminate the room. In looking around our individual corners, we share the assumption that there *is* in fact a light switch in the basement. We know that finding it is not equivalent to knowing how to operate it, and even when we know how to operate it, not all of the basement's nooks and crannies will be illuminated. Despite acknowledging these limits, we are not content with relativistic story-telling that keep us in the dark. Social science, though not a linear road towards total disclosure, provides a method of pushing the limits of what we know and understand, and a means for discerning the next most pressing set of questions (Rescher, 1998). Research, in the broad sense of the word (see Byrne, 2001), is our principle means of shifting our veil of ignorance.

The line of reasoning behind this worldview is decidedly realist. For starters, we assume that there *is* a reality that can be *known* - even though the proverbial toolbox to uncover that reality is less than perfect. Researching complexity is difficult, time-consuming, confusing and sometimes vague. But that does not mean that we should not try to do it. As Sanderson (2000) remarks, methods that don't account for the complexity of policy systems impede policy learning and evaluation. The aim of this chapter is to demonstrate why researching the complexity of public decision-making requires a realist stance and to show how the comparison of qualitative in-depth case studies offers a concrete method for moving from a philosophical stance to one of practical scientific inquiry.

We will first revisit the argument from Chapter 1 that reality is complex (Section 6.2.1) to remind ourselves that complexity is not a truism, and to make the case for critical realism. Four properties of complex reality that must be taken into account when doing research are then described. Next, the case for a critical realist ontology is presented (Section 6.2.2). To move from this ontology to a concrete methodology for collecting and analyzing data, we define two focal points for structuring research in complex systems: context and time (Section 6.2.3). We will then discuss the need to research complex systems as case studies, by focusing on the specific method of Qualitative Comparative Analysis (QCA), first proposed by Charles Ragin in 1987 (Section 6.3.1). The discussion on the operation of QCA is based on Wuisman's cycle of discovery (2005) and consists of two parts: zooming in on richly detailed in-depth case studies (Section 6.3.2), and zooming out to compare the in-depth case studies to arrive at statements about the trajectories of complex systems (Section 6.3.3). Although QCA is a complexity-informed method, it has a number of disadvantages that need to be addressed (Section 6.3.4). The chapter concludes with a reflection on causality and general laws (Section 6.4.1) and several key observations (Section 6.4.2).

6.2.1 REAL COMPLEXITY, COMPLEX REALITY

The book's stance on public decision-making being a complex endeavor is based on the idea that complexity is a real, non-constructed, property of the social world. This is an axiomatic statement about how the world works, and further proof of it is logically impossible. The most fundamental statement about complexity can be found in Chapters 1 and 2 and concerns the openness of complex systems, i.e. the fact that the world is composed of open systems, which in turn have other open systems nested within them (Byrne, 2005). The openness of systems means that they interact with other open systems in their environment (Cilliers, 2001). To be more precise, systems are situated in a particular context that brings about systemic changes as environmental influences are internalized through interaction, and can become part of the system's structure.

Emergence, which was discussed in Chapter 4, aids our understanding of this process because it emphasises how systemic structures and properties emerge out of local interaction. Structure does not exist without processes of interaction. Without interaction only an assemblage of separate components is visible. We follow Reed and Harvey (1992) and postulate that the real world consists of open systems which are emergently-structured. Note that emergence does not indicate a discrete entity or measurable phenomenon (Elder-Vass, 2005). Rather, it serves as an ontological heuristic device for thinking about the nature of complex causation, i.e. it helps us understand how reality unfolds. Consequently, we need to remind ourselves that if social reality is the non-linear emerging result of interacting components, it follows that the future is not the mirror of the past, as summarized in the argument for a developmentally open world and time-asymmetry (see Chapter 4).

The argument above summarizes the core aspects discussed in the previous chapters and leaves us with four fundamental properties of reality, which in turn lead to four conclusive statements about how this reality can be understood (cf. Byrne, 2011b; Gerrits & Verweij, 2012). First, if reality emerges through interaction, it is non-decomposable. Explaining a particular component of a system as a discrete entity does not explain the whole since it is the internal and external interactions that give rise to real structures and processes. Second, if complex systems are open and interact with their environment, their reality is situated or nested such that each instance of a particular system has to be understood as a specific conjunction of factors. Explanations of a particular phenomenon are therefore contingent: temporal in time and local in place. Third, these first two aspects mean that reality cannot be compressed without losing some of its aspects. Reduction or compression is inevitable in research because of practical constraints and is inherent to human cognition, as discussed before. While any explanation or modelling of complexity is therefore reductionist by definition (Cilliers, 2002; 2005a; 2005b), there is a real difference between reductionism as a consequence of practical limitations, and reductionism as expressed in Occam's Razor. We (grudgingly) accept the first but try to avoid the latter because that would put us on the wrong track. Fourth, time-asymmetry means that the past is not a perfect predictor of the future. The past can inform us about what may come, but a developmentally open world introduces randomness and uncertainty. These four properties and statements about how reality can be understood are the foundations for this chapter.

6.2.2 A CRITICAL REALIST ONTOLOGY

Complex reality, as summarized by its four properties, implies that reality is driven by causality, which can be researched and known. This, in turn could imply a positivist ontology. Some early attempts at formulating the mechanisms of complexity, such as the work of Axelrod or Holland, are clearly rooted in such a positivist ontology, even though those authors stressed the importance of anti-reductionism. At first glance, the complexity sciences appear ambiguous about the ontology and epistemology of researching complexity. On the one hand, they have inherited positivism from the physical sciences that have theorized heavily about complex systems; however, on the other hand, this positivist stance has been criticized and amended when used to analyze *social* complexity (Byrne, 2002; 2005; 2011a; 2011b). The fact-value dichotomy that underlies positivism has been thoroughly undermined in the realm of social sciences (Fischer & Forester, 1987; 1993; Fischer, 1998; 2003) and analysing and understanding social complexity implies a stance that acknowledges that reality cannot be wholly and unambiguously understood through the deployment of research tools (Cilliers, 1998). As argued in the previous chapters, human interpretation and the social construction of meaning play a pivotal role in understanding complexity and its causality. This makes a strong case for the adoption of a non-positivist approach when analyzing the complexity of public decision-making (e.g. Haynes, 2001). Although taking such a non-positivist stance allows for the convergence of fact and value in the analysis of complex causation and acknowledges the locality of the knowledge that is generated, it does not go as far as to accept postmodernism and its relativism. Instead, it assumes that explanation is possible, as long as such explanations are understood to be contingent, meaning they are temporal in time and local in place (Byrne, 2005) even though temporal cause and effect relations do exist and sometimes can be known. The ontological point of departure is therefore neither positivism nor post-positivism but a form of complex realism that mediates between the two (Reed & Harvey, 1992; Byrne, 2002).

The above echoes Kant's stance on empirical realism, that reality exists independently, outside our knowledge and perception, and can only be accessed through human agency (Losch, 2009). The view also goes beyond Kant's in that it assumes that scientific research is a method that can *approach* reality through the deployment of scientific tools (Sayer, 1992; 2000). This way of thinking is propagated in critical realism, which may be a key avenue for developing the relationship between empiricism and the interpretive sciences (Wuisman, 2005). Critical realism has been adopted in many scientific domains (Easton, 2010), although there are multiple and sometimes disconnected strands that fall under its banner (Losch, 2009), which can be confusing. We specifically build our argument on the work by Roy Bhaskar (e.g. 1979; 1988; 2008). Bhaskar originally placed his work under the label of transcendental realism (Losch, 2009), but over time, it came to be known under the nomenclature of critical realism, which we use as shorthand for Bhaskar's philosophy and related accounts (cf. Gerrits & Verweij, 2012).

Central to Bhaskar's argument for critical realism is the existence of a reality independent from human observation and cognition. However, contrary to similar positions in the social sciences, he believes that we can observe, research and analyse the mechanisms underlying the occurrence of such events. Thus, Bhaskar moves beyond the idea that humans can only describe what they think they perceive without claiming to gain better access to the causal mechanisms behind events, processes or human behaviour. To structure this argument, he presents a stratified perspective on social reality to give meaning to how one can understand the world. As Sayer (2000: 11) puts it, "critical realism distinguishes not only between the world and our experience of it, but between the real, the actual and the empirical". The empirical is the domain of personal experiences. Aside from these experiences, there are the realities of actual events, processes and behaviour, and the mechanisms underlying these events and processes that take the form of structures and powers. The key point here is that these mechanisms exist a priori, and a particular configuration may or may not bring the mechanisms into action. A well-known and oft-cited example of how this works is labour power. The power to perform labour is very real, but it becomes actualised at the moment labour is actually performed (cf. Sayer, 2000). Thus, in Bhaskar's view, external events, processes or behaviours (second stratification) and the mechanisms behind it (third dimension) can be observed (first dimension) and therefore researched as they unfold (Easton, 2010).

Here, research means tracing the causality of what is being observed (Hammersley, 2008; 2009). More precisely, this means generating preliminary reconstructions of how events follow from previous events, or how processes develop through time and by what mechanisms human behaviour comes about. Causality implies that something is made to happen, that there is something that produces, generates, creates or determines something else, or "more weakly, what 'enables' or 'leads to' it" (Sayer, 1992: 104). However, causality is more than just trying to establish what leads to what. First, Williams (2009; 2011) argues that while there are often sufficient conditions for something to occur, "but they are not necessary ones, except under specific circumstances whereby the 'natural necessity' is actualized." (Williams, 2009: 3). In open complex systems, the same causal power can produce different outcomes, whilst it is entirely possible that different causal mechanisms can produce the ostensibly same or similar results (Sayer, 2000). Thus, it is logically impossible to know 'necessity' up to the moment it is actualized. Moreover, any explanation of causality is partial and contingent because reality comes about through the conjunction of certain factors. Some causal mechanisms operate and some do not, which marks the difference between potential and achieved influences, but does not mark an apparent difference between real influences (realized, therefore observed and appearing to be present) or unreal influences (unrealized, therefore unobserved, and appearing to be absent). This difference between potential and realized influence can differ between events, processes, systems or cases. Whether certain mechanisms are in operation or not is an empirical question.

We have to constantly remind ourselves of the limitations of thinking in terms of real causality. As pointed out by Cilliers (e.g. 2001; 2005b) and others, it is impossible to uncover reality unambiguously and comprehensively. Because we cannot research the whole world, each research method inevitably represents a choice to focus on some aspects whilst ignoring others. Individuals nevertheless act upon their interpretation of reality, as shown in Chapter 4, and this reality becomes real as a consequence (Easton, 2000). It is for this reason that human agency (primarily human perceptions and actions) should be central to any analysis of social complexity, as will be further demonstrated in Section 6.3.2.

Open systems also imply that one cannot control the conditions under which something occurs. The situated or nested nature of complexity implies that conditions will differ from one situation to the other and that the results that are observed could be generated through different specific configurations of conditional variables. The absence of control over the conditions in the real world (as opposed to the closed world of lab-based experiments) means that it is impossible to establish a definitive causal account, because one does not know how the conditions in which a certain event or process takes place impact causality, i.e. whether they promote or dampen the occurrence. Thus, we take issues with the golden standard of the double-blind experiment and its goal to establish causal relations *despite* uncontrollable real conditions (cf. Levitt & List, (2007); Callaghan, 2008). Differing conditions are not simply 'noise'; they are explanations for how something comes about.

What does critical realism then tell us about the possibilities of prediction by means of analysing complex systems empirically? A developmentally open world always has a degree of randomness (Prigogine, 1997) and Bhaskar's theory is built on the same premise. When observed events, processes or behaviours are not caused by a supposedly fixed set of variables but rather by a conjunction of variables or conditions at a certain point in time, what is observed can only be explained retrospectively, and not predicted (Williams, 2009). It should be emphasized that this does not imply total unpredictability since we can in fact learn from past experiences and from research that shows how things came about. We could, for example, show that some future system states are more probable than others given a certain configuration of conditions, (Byrne, 2011b). However, it has to be understood that such estimations shouldn't be built on the homogenizing assumptions of time-symmetry.

The argument above can be summarized in a description of the nature of complex causality, following Byrne (2011a; 2011b). Complex causality is an interaction of generative mechanisms with specific contexts, resulting in unidirectional outcomes, meaning that the outcomes are subject to time-asymmetry. In addition to Byrne's description, we argue for the plural form of the word 'mechanism' because multiple mechanisms produce an outcome in conjunction with contexts. In other words, a space of possible combinations exists, from which a specific configuration is triggered at a given point in time. Naturally, the next question then is: how can the complexity of public decision-making be researched?

6.2.3 FOCAL POINTS: CONTEXT AND TIME

It is important to consider the implications of the four main properties that inform our critical realist ontology for researching complexity. While the four properties are rooted in the topics discussed in the previous chapters, i.e. in the systemic (non-decomposable, conjunctive, non-compressionable) and the dynamic (time-asymmetric) nature of complexity, for the sake of convenience, in this section, we group those issues under the headers of context and time.

Context refers to the fact that each instance of an open and nested system forms an assembly of generic elements or properties (i.e. something that appears elsewhere in similar cases) and specific elements or properties (i.e. something that only occurs in this instance). In short, we deal with situated complexity when analyzing public decision-making (Buijs, Eshuis & Byrne, 2009). For instance, we could consider the urban environment as a situated complex system (cf. Batty, 2010; Roo, Hillier, Van Wezemael; 2012; Roo & Schwartz, 2001; Zuidema &

Roo, 2004) and observe similar patterns of development across many West European cities. An example of such a pattern is that the manufacturing sector tends to move outwards from cities, to be replaced by tertiary sector companies. However, other developments are very specific to a particular city. For example, compare the ways in which Rotterdam, Coventry and Dresden were reconstructed following their destruction during the Second World War. Over time, the interaction of generic and specific elements leads to the creation of a local situation that is unique, even though it retains recognizable elements (Byrne, 1998; 2001; 2003; 2005; Marshall, 2009). Research into the trajectories of restructuring industries in post-socialist Łódz (Poland) by Dornish (2002) demonstrates how shifts in those trajectories occurred because of specific changes in the local configuration of those industries. These examples draws attention to the fact that causal relationships are case-specific by definition, and that local conditions consist of explanatory variables for the whole that is being observed (cf. Mjøset, 2009). It means also that open complex systems do not operate exclusively according to general rules occurring in all contexts (Buijs, Eshuis & Byrne, 2009: 37), or that they are purely idiosyncratic. A systematic comparison can reveal both differences and similarities between the operations of different systems (Verweij & Gerrits, 2012).

Time, the second issue here, refers to the fact that complex systems are dynamic: they are developmentally open and change over time. A longitudinal analysis is required to explain how a complex system comes about and evolves over time. The non-linear dynamics of such systems mean that periods of relative stability can be punctuated with periods of relatively swift change, during which events follow each other more rapidly than before or afterwards. The case of the Unterelbe discussed in Chapter 3 is a good example of such erratic change. The risk of using comparative statics by taking snapshots of data at fixed intervals is that the non-linear nature of such systems will not be noticed and the alternating periods of stability and swift change will be missed (Gerrits, 2008). A continuous longitudinal approach should also be very detailed as the non-linear emergence of structures and processes cannot be deduced in a linear way. Inevitably, this means that a high resolution view of past developments is necessary to find the causal relationships that exist between events and processes.

This reduces the risk of overlooking important developments that could have had an effect on future changes.

In taking context and time as the points of reference for social scientific research, how does one conceive of complex systems? Byrne argues that cases are the methodological equivalent of complex systems; or, alternatively, complex systems are cases and therefore should be studied as such. (2011a). Viewing complex systems as cases allows one to follow a particular trajectory through time, bearing in mind that such trajectories are driven by a combination of recurring patterns and local conditions, both of which can only be identified through comparison (George & Bennett, 2005). But cross-case comparative studies often fall short in accounting for the rich details of local conditions that are important in shaping each case. Thus, we need to find a research method or tool that enables us to move back and forth between highly detailed in-depth case studies and comparison between multiple cases, and that also helps us account for the dynamics of complex systems (Vesterby, 2008). That method is offered in the shape of Qualitative Comparative Analysis or QCA.

6.3.1 QUALITATIVE COMPARATIVE ANALYSIS

Qualitative Comparative Analysis has been proposed, most notably by David Byrne (2005; 2009; 2011a; 2011b) and Charles Ragin (1987; 2000; 2008) as a complexity-informed research method that mediates between the understanding of complexity and a knowledge of generality (Ragin, Shulman, Weinberg & Gran,2003), which is very useful in analyzing and evaluating policies (Befani, Ledermann & Sager, 2007; Blackman, 2006; Byrne, 2011a; Verweij & Gerrits, 2012). QCA is a method that allows the researcher to explore a particular case, to generate preliminary explanations from that case, to test those explanations against a larger (but limited) set of cases, and to use the results of such a test to improve our understanding of singular cases. It does this by preserving a considerable level of complex details, and as such can be used to "achieve a systematic comparison across a smaller number of individual cases (e.g. a sample of between 10 and 30 cases) in order to preserve complexity, and yet being as parsimonious as possible and illuminating otherwise often hidden causal paths on a micro level" (Rihoux & Lobe, 2009: 228). Thus, its main property is that it is able to account for the context of a social phenomenon.

Unlike other methods that force a choice between the in-depth case studies and comparative studies, QCA also allows for an exploration of causal patterns over different cases. It is important to note that in QCA, variables are reframed as causal conditions or sets (Byrne, 2011a) because the framework does not agree with the conventional assumption that each individual variable "has an independent impact on the outcome" (Ragin, 2000: 15). Instead, it treats causes as combinations that bring forth outcomes. As a comparative case-based approach, it then allows for the examination of multiple causal configurations (Byrne, 2009), i.e. it helps uncover how combinations of causal conditions produce outcomes. Consequently, it can also help identify how apparently similar outcomes arise out of different combinations. This in turn implies that certain factors can generate different effects in different contexts. These three characteristics of complex causality are referred to as conjunctural causation, equifinality and multifinality (Grofman & Schneider, 2009; Schneider & Wagemann, 2010; Wagemann, 2009).

QCA mediates between in-depth case studies and case comparison by iterating between induction and deduction in a single research cycle. This approach suits the goal of researching complexity from a critical realist ontology particularly well (cf. Wuisman, 2005). As with any method, QCA compresses data but remains committed towards preserving the essentially qualitative and detailed nature of the data in order to facilitate a meaningful comparison across cases. While other comparative methods are often geared towards a reduction in environmental influences in order to identify the key variable(s) that control the phenomenon, QCA is situational by definition. It helps in the identification of how both generic patterns and idiosyncrasies generate a case. In addition, QCA can be used to describe the trajectories of systems (Byrne, 2011c). It works as a descriptive tool that addresses the configurations of elements characterizing a certain system state at different moments in time (e.g. Vis, 2007). This addresses the nature of complex systems as heterogeneous cases made up of different conditions, enabling it to account for the emergent nature of complex systems (Ragin, 1987). However, it focuses more on context and the nested character of cases, and less on time, an issue we will return to in Section 6.3.4. The characteristics of QCA are summarized in Table 6.1 and compared with two archetypes in social scientific research, namely variable-oriented and case-oriented research.

	Requirements:			
Approach:	In-depth vs. generalization	Case-based vs. comparative	Attention to context	Attention to time
Variable-oriented	Generic patterns	Comparative, not case-based	Limited	Limited
Case-oriented	In-depth, focus on idiosyncrasies	Not comparative, case-based	Yes	Yes
Qualitative Comparative Analysis	Iterations between generic patterns and idiosyncrasies	Case-based comparison as main feature	Yes, strong	Yes, but weak

Table 6.1: QCA as a complexity-informed method in social scientific inquiry. Adapted from Verweij & Gerrits, 2012.

QCA meets most but not all of the four properties of complexity as formulated in Section 6.2.1. It builds on the notions of conjunction and time-asymmetry. It also aims to explain cases as a whole and as such addresses the non-decomposable nature of reality. Inevitably, as with any method, applying QCA requires that boundaries be drawn around the subject, which compromises the property of non-decomposability and leads to compression. However, the decomposition and compression that takes place in the research cycle can be made transparent and accountable, as shown in the following sections where we discuss the steps to be taken for using QCA.

6.3.2 ZOOMING IN: UNDERSTANDING IN-DEPTH CASE STUDIES

Complexity is notorious for being recursive to the extent that closer looks reveal more and more details that mirror the whole, but not in such a way that the whole can be fully known by merely looking at the component properties (cf. Cilliers, 1998).

A single in-depth case study is a study of the intricate workings and details of a particular complex system, and allows the development of thick case descriptions. Case studies can be approached from multiple angles. For example, in positivist approaches, the researcher informs himself theoretically prior to the empirical stage and this differs primarily with a grounded theory approach in which the researcher develops a preliminary theory based on what he has seen, heard or read during the empirical stage (Blatter & Blume, 2008). A complexity-based approach does both at different stages because it is theory-informed, yet it accepts that no theoretical model can be accurate enough to fully capture a particular case. .

To unearth something meaningful, a starting point must be identified, and a general objective, such as tracing path dependency in administrative cluttering, or mapping the co-evolution of a set of natural and social systems, is helpful in this case. But before we lose ourselves in elaborate theoretical models, we need to remember Checkland's lessons (Chapter 2). His argument for science in action points at the impossibility of the researcher's task of fully defining a complex system and its workings prior to empirical research. A theory or a concept can help as a starting point, but the researcher should not expect to be able to develop an accurate model beforehand.

Checkland (1981) and others, such as Uprichard & Byrne (2006) and Wagenaar (e.g. 2007; Wagenaar & Cook, 2003), argue that we can only learn about the constitution and operation of a complex system through the eyes and actions of people working in and with complex systems (public decision-makers in our case). Such people are the proverbial spies that help the researcher disclose and understand perceptions, actions and responses. Such individuals are pivotal in two ways: they guide the researcher in generating temporal system boundaries, and preliminary causal relationships (Byrne, 2011a; 2011b). Both aspects help account for the local conditions that bring about the properties of the system.

With regards to the first aspect, the boundaries of the system are constructed by actors who act consequently to it as explained in Chapters 2 and 4. It is therefore important for the researcher to understand the social construction of system boundaries, i.e. how system boundaries emerge and evolve through an interaction between actors (cf. Midgley,

Munlo, & Brown, 1998; Ulrich, 2005). This in turn allows for the reconstruction of how boundary judgements converge, diverge or intersect and how this influences the processes of public decision-making. Note that the goal is not to develop unambiguous system boundary judgments, but to reconstruct the formation of such boundaries, and understand how elements from the environment are incorporated into or kept out of the system (cf. Cilliers, 2001). Boundary judgments between ostensibly different system types may seem obvious, but not always. For example, Williams (2009) and Gerrits (2008) argue that social and physical objects are not clearly separable because the physical world is interpreted by actors, and the boundaries that are drawn differ between such actors.

This points to the second aspect, namely the fact that the causal relationships that generate a system are too numerous and too intricate for a researcher to determine without the help of actors in that system (Tsoukas & Hatch, 2001). Essentially, actors develop and use mini-theories about how a system works to understand what they will need to do to achieve their goals (Wagenaar, 2007). For example, they reason that "if I decide to do this, that will be the consequences", or "that situation asks for this decision to be taken." These are all mini theories, or preliminary causal reasoning about the way particular segments of the system work, and how these segments could be influenced. Since actors are central to social systems, the way they think and act is what makes the system tick. Wagenaar's study of the decline and revival of neighbourhoods (2007) provides a clear example. He notes how residents and others "[...] have a ready understanding of the complexity of the issues that affect them, although they do not use the analytical lingua franca of the academic experts [...]" (2007: 26). Although the explanations offered by respondents are partial and ambiguous, this is not seen as a weakness because "the precise relationship among [...] elements of the neighbourhood life *is* unclear. Nobody, not even policy experts or city planners, is able to specify the relationship among these variables [...]" (2007: 27; emphasis LG). Thus, Wagenaar moves from narratives to the reconstruction of the way the neighbourhood works as a complex system, and in doing so, assembles those mini-theories into a systemic whole.

On the operational level, these two aspects demand that the researcher abandon surveys in favor of open or semi-structured interviews and participant observations. Data from both sources should be open coded in order to exhaustively map the various occurrences, events, processes and behavior, and then recoded to improve its accuracy. Coding allows bits of data to be systematically and systemically related, such that one is found to 'be part of the other', 'occurs in the context of', 'define this', 'contradict that' and so on. Over a number of interviews and observations, the codes will be grouped and arranged so that they reveal the workings of a system. While coding itself cannot generate interpretations of the data, it is a valuable aide to the researcher in the process of interpretation.

Researchers have to consider the level of communication of the respondents when coding and interpreting the data. In developing their mini-theories, individuals rely on their past experiences and expectations (e.g. Weiss, 1995). Consequently, action within the system is both responsive and constitutive (Uprichard & Byrne, 2006). On top of that, respondents also aspire to reach certain favorable situations, and this desire intermingles with their memories of the past, current experiences and future projections (Wagenaar & Cook, 2003). Therefore, while the data is used to reconstruct or map the system, it also makes the researcher and the respondent reflect upon the process of reconstruction.

In this way, as Checkland (1981) and Flood (1999a; 1999b) point out, reflexivity becomes an operational component of complex social systems (Leydesdorff, 1997) and recording, coding and relating qualitative data becomes science in action. For this reason, some authors regard reflexivity, participatory observation, and even deliberate action as an important part of the complexity researcher's toolbox (Checkland, 1981; Checkland & Holwell, 1998). By adding deliberate action, the complexity researcher moves into the realm of action research (cf. Burns, 2007).

For example, Byrne (2011a) recounts how in his work as a researcher at the North Tyneside Community Development Project, research was co-produced with community members who had a vast memory of the area's evolution. Consequently, the researchers learned that both the history of the area and the existing relationships within the housing system were crucial in identifying how the present state came to be and what future states were possible with the implementation of specific policies. Arguably, such knowledge could only be harnessed by being actively engaged with the project, not only to collect data but also to try out policy alternatives. Byrne's argument is that action research provides direct access to the workings of complex systems. More pragmatically, action research can provide a shortcut to knowledge because all that matters is not the mapping of all possible states but rather of *what works*, i.e. what has brought about the desired outcomes. More fundamentally, research results are always neater than the messy reality of public decision-making and it can be argued that it is imperative that the researcher *experiences* this messiness to understand that what works on paper does not necessarily work in practice. Thus, there are ample reasons for the complexity researcher to be engaged with her subjects, both for deeper scientific understanding and better policy information.

The 'Rondom Arnemuiden' case study (Schie, Edelenbos & Gerrits, 2010; Schie, 2011) is a useful example. This case concerns the regeneration of a small and decaying village south of Rotterdam, the Netherlands, by digging and dredging a water body from the Veerse Meer Lake nearby towards the village. The municipality envisioned that this would enhance the attractiveness of the area, making the construction of middle- and upper range housing for the well-to-do more feasible, consequently leading to an increase in the value of the land. In addition, it would increase the proportion of land designated for nature conservation, attracting more tourists. The municipality hired an urban planner, renowned for her successful regeneration projects on the south banks of Rotterdam, to develop the plan. Although the policy plan looked attractive and feasible on paper, implementation failed because the locals were resolutely against any plan that involved the breaching of dykes and inundation of their polders.

The authors of the case study were hired to provide advice as to how this implementation failure could be rectified, and they adopted an action research approach to understand the mechanisms behind this failure. They interviewed local residents, farmers and entrepreneurs, as well as government officials assigned to the project. They also observed how these parties interacted during formal and informal meetings, and analyzed policy documents. The results were discussed during a number of workshops with both civilians and government officials. These workshops helped the researchers understand the meaning of the findings, and helped the participants understand their situation, and to better assess possible solutions.

The researchers found that this ostensibly simple project was so deeply systemic that it paralyzed decision-making. A key factor was a history of distrust between villagers and authorities, to the extent that some officials were afraid of visiting the village. There were also numerous substantive issues that stalled decision-making. For example, attracting more tourists required an upgrade of the existing road as per regulations, which in turn required a revision of the access routes to the main highway. Since each highway was only allowed a limited number of access routes, this realignment in turn required a complete revision of all access and exit points in the whole region. Other issues, such as water management or property development, turned out to be equally systemic.

The 'simple' plan then propelled itself into a complex system. The most prominent example appeared when the municipal authorities kept establishing new working groups to mirror the substantive complexity. They ended up with more than ten separate working groups over the course of two years, including a working group for coordinating the other working groups! Mapping the boundaries and interlocking effects using the qualitative data obtained during the process proved vital in understanding this case. This understanding was achieved by going into the field, experiencing what the actors did, and how they constructed and reconstructed the way their project worked. For example, the proliferation of organizational control mechanisms in the shape of the municipal working groups, or the inability to upgrade a simple road, may have seemed to be due to the lack of quality of the officials. The data showed that the responses of those actors were entirely built on a rigid logic, even though most people regretted the outcome. The researchers came up with an evolving model of this case as a complex system. This model delivered useful insights but little else beyond that, as is common with single case studies. To what extent was this policy implementation failure due to a unique conjunction of conditions? Alternatively, was this a recurring pattern of events? Answering such questions requires comparing the data from different case studies.

6.3.3

ZOOMING OUT: COMPARING CASE STUDIES IN FUZZY SETS

Suppose that we use the data from one complex system in the shape of a single case study to build a temporal model of how that particular system works. Ideally, we would like to understand the extent to which specific local conditions generated the observed effects in combination with generic patterns that reoccur in other cases. QCA offers an approach that allows a limited number of in-depth case studies to be compared by treating them as sets. The findings from individual case studies can be grouped into a set; for instance, an 'implementation failure' set could be created from the 'Rondom Arnemuiden' case study if we were interested in comparing its outcomes with other urban regeneration attempts that also failed at the implementation stage.

The construction of such sets can be rooted in theoretical expectations about certain patterns of interaction between variables (for example, the expectation that distrust between actors leads to implementation failure), or generated from empirical data, as described in the previous section. Such sets can be *intersected* using the operator 'logical and' (this refers to conjunctural causation and is indicated with a '*') or *joined* using the operator 'logical or' (this refers to equifinality and multifinality and is indicated with a '+'). QCA enables researchers to systematically compare and analyze the conjunctions of sets (Ragin, 1987; 2000; Smithson & Verkuilen, 2006), and shifts our focus from *cor*relations to *set* relations. That is, instead of studying the net-additive effects of variables, QCA studies the necessity and sufficiency of relations between (combinations of) sets and outcomes (Ragin, 1987; Schneider & Wagemann, 2010).

Using sets to analyze cases can help reveal how certain outcomes come about in public decision-making (Verweij & Gerrits, 2012). A certain condition is considered necessary if it has to be present for the outcome to occur, and this is indicated by the statement: '*A*' being a subset of the condition '*B*'. That is, every case in the sample that exhibits *B* also exhibits *A*; if the case does not exhibit *A*, then it cannot be in set *B*. A condition is sufficient if it can produce the outcome by itself, and this is indicated by the condition being a subset of the outcome. However, Ragin (2000) argues that, in reality, there are no pure necessary or sufficient conditions to explain an outcome. However, certain conditions can be necessary in certain contexts, as captured in the concept of *configurations*. Such conditions are called INUS conditions, which are "*insufficient* but *non-redundant* part of an *unnecessary* but *sufficient* condition" (Mackie, 1980: 62). For example, imagine that the Rondom Arnemuiden case study c^1 is compared to two other cases $c^{2\text{-}3}$ that have different scores on three causal conditions *A*, *B*, and *C*. Now suppose that the Rondom Arnemuiden case exhibits conditions *A* and *B*, the second case conditions *B* and *C*, and the third case conditions *A* and *C*. We are then left with three intersected sets: *A**B*, *B**C*, and *A**C*. Together, these three different paths produce a certain outcome, e.g. the outcome 'implementation failure'. For convenience, this can be written as a Boolean expression:

$$\underbrace{A^*B}_{(c^1)} + \underbrace{B^*C}_{(c^2)} + \underbrace{A^*C}_{(c^3)} \rightarrow \text{implementation failure}$$

The expression captures the argument that *A* and *B*, or *B* and *C*, or *A* and *C* result in policy implementation failure. It shows that none of the three conditions *A*, *B*, or *C* is sufficient by itself to bring about the outcome 'implementation failure'. It also shows that none of the three conditions is necessary. For instance, condition *A* is not necessary since the outcome can appear with the combination *B**C*. This means that *A* is an INUS condition: it is an insufficient (i.e. it cannot produce the outcome by itself) but non-redundant (i.e. it is a necessary condition in both the combinations *A**B* and *A**C*) part of an unnecessary (i.e. *A**B* and *A**C* are not necessary since implementation failure also appears in *B**C*) but sufficient (i.e. *A**B* and *A**C* are sufficient for implementation failure to occur) condition. Note that Ragin (2008) states that QCA allows the conditions that are found to be supplemented with theoretical and substantive knowledge to improve their explanatory power.

Building and relating sets to understand how certain outcomes are generated in different cases lies at the core of QCA. However, the method presented above is somewhat crude as it can only operate with dichotomous values, i.e. whether a causal condition is either present (1) or absent (0). Such a dichotomy can be considered an oversimplification of social reality. Ragin overcame this limitation by developing an improved method called fuzzy-set QCA or fsQCA (2000; Rihoux & Ragin, 2009). This method allows for a gradient rather than a dichotomy in the degree of set membership, i.e. conditions could be assigned membership values between 1 and 0. Thus, a factor does not need to be fully present or absent in a particular case, allowing for more nuanced distinctions. Note that a value between 1 and 0 does not mean that the condition is a continuous variable. Rather, it allows the researchers to assign set membership by taking into account the different combinations of conditions that lead to a certain outcome.

The 'Rondom Arnemuiden' case study was used to demonstrate the 'implementation failure' outcome, and it was compared with two other imaginary cases with the same outcome. However, the word 'same' can be deceiving here, because such clear-cut outcomes are rare in real-life public sector decision-making. While a policy may not be implemented at all in one case, it may be partly implemented in other cases. For instance, the Rondom Arnemuiden case showed that formal implementation failed. However, some of the government officials from the water board had begun to rework their policies to include the construction of a new water body, and some farmers were planning to sell land to the government so that it could start developing houses. Therefore, while official implementation failed, some of its aspects became real through the anticipatory actions of certain actors. This is not an exception, because other cases are also very likely to have witnessed similar dynamics. In-depth case studies can reveal the extent and causes of such differences, and fuzzy-set QCA takes these differences into account when the cases are compared.

The essence of fsQCA is that we want to express and retain the detailed conditions that lead to each outcome. For example, we may conclude that the reasons for and extent of 'implementation failure' is greater in c^1 than it is in c^2. Thus, c^1 is a more prominent member of the set 'implementation failure' than c^2. This distinction is formalized by assigning a membership score of 1.0 to c^1 and a membership score of 0.75 to c^2. Consequently, the set 'implementation failure' constitutes a 'difference-in-kind'. Once each case (c^{1-3}) has been assigned a membership score on each of the conditions, the researcher can proceed to comparing them using fuzzy sets.

The next step in that process is the construction of a truth table (e.g. Ragin, 1999). A truth table lists all logically possible combinations of causal conditions, and sorts the cases according to these conditions. That means assessing and ranking the presence of combinations that have been observed empirically. Each row in the table presents a unique (theoretical) configuration of intersected sets. An output value needs to be defined for each possible combination. This output value is based "on the scores of the cases which share that combination of scores on the independent variables. Thus, both the different combinations of input values (independent variables) and their associated output values (the dependent variable) are summarized in a truth table. [...] each row is not a single case but a summary of all the cases with a certain combination of input values" (Ragin, 1987: 87). Examples of truth tables using fuzzy sets can be found in Ragin (2000).

Note how the construction of the truth table moves us from rich, detailed but singular case studies towards the reduction necessary for systematic comparison. The next step is to minimize the truth table to its essential core to generate a statement about the occurrence of patterns across the cases in the sample, i.e. the so-called solution. This minimization is structured using Boolean algebra. The basic procedure is summarized as follows: "if two Boolean expressions differ in only one causal condition yet produce the same outcome, then the causal condition that distinguishes the two expressions can be considered irrelevant and can be removed to create a simpler, combined expression" (Ragin, 1987: 93). The absence of a condition is indicated using the '~' sign. For instance, a truth table of five hypothetical cases could generate the following Boolean expressions:

$$\overbrace{A*B*C}^{(c^1)} + \overbrace{A*B*{\sim}C}^{(c^2)} + \overbrace{A*{\sim}B*C}^{(c^3)} + \overbrace{A*{\sim}B*{\sim}C}^{(c^4)} + \overbrace{{\sim}A*B*C}^{(c^5)} \rightarrow \text{implementation failure}$$

The expression depicts how five different empirically-observed paths lead to implementation failure. The next step is to identify the sufficient, necessary and/or INUS conditions. For example, the expressions from the first two cases can be minimized to *A*B* because implementation failure occurs regardless of whether condition *C* is present. The expressions from c^1 and c^4 can be further minimized to *B*C*, and so on. The following solution formula results:

$A + B^*C \rightarrow$ implementation failure

The solution formula is a statement about the patterns and conditions leading to the outcome 'implementation failure' across cases. It is a minimized and somewhat precise indicator for the patterns that occur across cases, but only makes sense if it is reinterpreted in the light of the individual case studies. Moving from in-depth cases, the Boolean minimization is only one part of the cycle of scientific discovery mentioned at the start of this chapter. It is pivotal that the researcher return to the in-depth case studies, and perhaps even add new ones, to understand the *meaning* of the solution formula. Adding new cases or new conditions may cause the solution formula to shift. This is only normal as the solution formula only extends to the cases being considered. It can give clues about what might occur in other systems in similar circumstances but it cannot predict what will actually happen in other systems in similar circumstances.

One might argue that the procedures of the truth table and Boolean minimization simplify the complexities of case studies too much. However, a formal comparison method does not only generate a solution formula, but also acts as a heuristic device that enables the researcher (and perhaps the respondents, in the case of action research) to reconsider the patterns observed. It also formalizes the procedure that researchers go through when researching multiple cases, even when comparison is not an explicit aim. In doing this, the essence of context for complex systems remains intact.

Note that this section tries to give an overview of a method in a very limited amount of space. There are many more issues surrounding fsQCA, such as counterfactual analysis (i.e. adding data that contradicts the findings from the initial set of cases) and consistency and coverage measures. More detailed overviews and summaries of the procedures can be found in Ragin (2000; 2005; 2008).

6.3.4

ISSUES WITH FSQCA

While fsQCA is a very promising method for analyzing the complexity of public decision-making, a number of weaknesses also need to be considered. Firstly, although it maintains the systems' integrity and contextual factors, it performs less well on time-asymmetry (the fourth property). In short: fsQCA is essentially static (Rihoux, 2003), and adding a time dimension would amount to a series of snapshots of system states over time (Rihoux, 2003; De Meur, Rihoux & Yamasaki, 2009). That may give an idea about the changes, or lack thereof, in the system state at any given *t*, but does not give much pertinent information about the mechanisms behind the transition (or lack thereof) between system states at t_1 and $t_{2\text{-}n}$. A work-around would be to include those mechanisms as a condition in the set, which requires the researcher to interpret which mechanisms contribute to changes between system states, and how. Another approach would be to complement QCA with other methods such as an analysis of sequences of events. Inevitably, each attempt requires a trade-off between researching context and researching time. This is inherent in the very nature of complex systems.

A second issue is the limited number of conditions that can be taken into account. Recall that the truth table contains all logically possible conditions. The addition of new conditions means that the truth table could extend exponentially, as does the addition of new cases. Still, such additions could lead to a more precise solution formula. One way of dealing with additional conditions is to go through the process multiple times (De Meur, Rihoux & Yamasaki, 2009). Each iteration will help the researcher identify which conditions matter and which conditions yield the same or similar results. These conditions can then be excluded from the analysis and respectively grouped together as macro-variables, freeing up space to add additional conditions.

Adding new cases is part of the philosophy behind QCA. The identification of the solution formula focuses our attention on the different causal routes towards a certain outcome. Consequently, the addition of new cases may lead to the discovery of new pathways. More common comparative methods are geared towards finding variables that control for differences and similarities in multiple cases. Such search is irrelevant when researching complexity, because each case has its unique pathway and comparisons should be used to highlight the particularities of the pathways. We need to remind ourselves that research is an ongoing cycle of discovery built on induction and deduction, and the solution formula is another, but not the final, step in that cycle.

6.4.1 CAUSALITY AND GENERAL LAWS

Discussing complexity research and the deployment of QCA leads us to revisit the question as to whether it is possible to assume, or perhaps even demonstrate, the existence of universal or general laws in social systems. General laws explain the true occurrence of something, regardless of time and place. Prediction is possible as a consequence. Such laws have four features that allow permanent explanation and prediction: [1] they must be rooted in empirical observations; [2] they must be applicable regardless of time and space; [3] they must represent the truth; [4] there must be natural necessity, i.e. the observations must not accidental (Mitchell: 2009). The real existence of such general laws is impeded by time-asymmetry, as argued by Prigogine (1997) and as shown by Mitchell in the case of biology. Non-ergodic chance events have a real impact on the trajectories of systems, obstructing pure repetition over time. This does not rule out laws as such, but means that the laws that are found are not general or universal by definition. Thus, while we may be able to explain or even predict, with various degrees of uncertainty, the trajectories of complex systems, we will fail to predict a specific trajectory to a singular case because of the elements of randomness and resulting non-ergodicity.

It is no surprise that social science has a hard time defining laws, let alone general laws. The scientific work-arounds for that problem were discussed in the previous chapters, such as the ceteris paribus clause and small world reasoning. Such approaches set clear limits and allow the development of general laws, as long as it is understood that the universal character of those laws depends on the limiting assumptions inherent to the use of the ceteris paribus clause or the construction of small worlds. In other words, the law is a derivate of the limits that were set. However true such laws appear, they are still not really general since they are conditional, i.e. they depend on the assumption that other circumstances do not matter, and we cannot fully satisfy ourselves with that.

One may ask whether the approaches and methods discussed in this chapter, despite the care taken to avoid positivist pitfalls, are still geared towards general laws. What does it mean to search for recurring patterns across cases? It is true that we try to seek how certain phenomena come about through the conjunction of local conditions and recurring patterns. It may even be possible that we can tell, within a certain degree of certainty, which factors cause a system to develop in a particular way. Still, this does not imply a search for general laws. The statements generated can never be extended beyond the set of cases, and while our approach may give pertinent information about a particular set of cases, it does not hold predictive power. Adding more cases may alter the recurring pattern found and the solution formula derived from it but that does not make it a general law. Although it is impossible to establish a definitive causal account, one should not give up trying to understand some of the observations that were made. QCA remains a realist method that, despite its limitations, can uncover a bit more of complex reality.

6.5 CONCLUSIONS

We started this chapter with the goal of explicating the ontology and epistemology behind the research findings and statements made in this book about the complexity of public decision-making. These findings and statements are rooted in a critical realist stance, which in turn informs the way research into the nature and operation of complex systems should be conducted. Following Byrne and Ragin, Qualitative Comparative Analysis was proposed as a promising research method. We discussed how the workings of a complex system could be described and analyzed via a single case study, and how multiple case studies could be used to trace local conditions and recurring patterns. Thus, a complexity researcher moves into a particular case as a participant observer or perhaps even an action researcher, collects data through interviews, observations, documents and actions, and then compares the findings against other cases using the procedures of QCA.

Although the argument here focuses on one particular method, this does not imply that it is the only method suitable for investigating complex systems. Approaches such as agent-based modeling also have much potential, modeling being especially interesting when it is used as a heuristic device for decision-makers working on real cases (Johnston, Kim & Ayyanger, 2007). The main point here is that researching complexity should not be a purely academic exercise, executed in the comfortable confinements of one's office or lab, especially not when this research is meant to inform public decision-making. Thus, we fully subscribe to Byrne's attitude that social science needs to be science in action. Public decision-making and the policies that result from it have a real impact on the real world, and that is enough reason to study it in its most natural, non-abstract way - complexity should be seen, touched, smelled and heard. Research into the complexity of public decision-making is ambiguous and by no means easy. While it may scare away the nihilistically romantic postmodernists or the stern formalists, such is the nature of complexity that it is imperative to understand it in full detail. I hope that this book has made a small and humble contribution to expanding that understanding.

ACKNOWLEDGMENTS, BIOGRAPHY, REFERENCES AND INDEX.

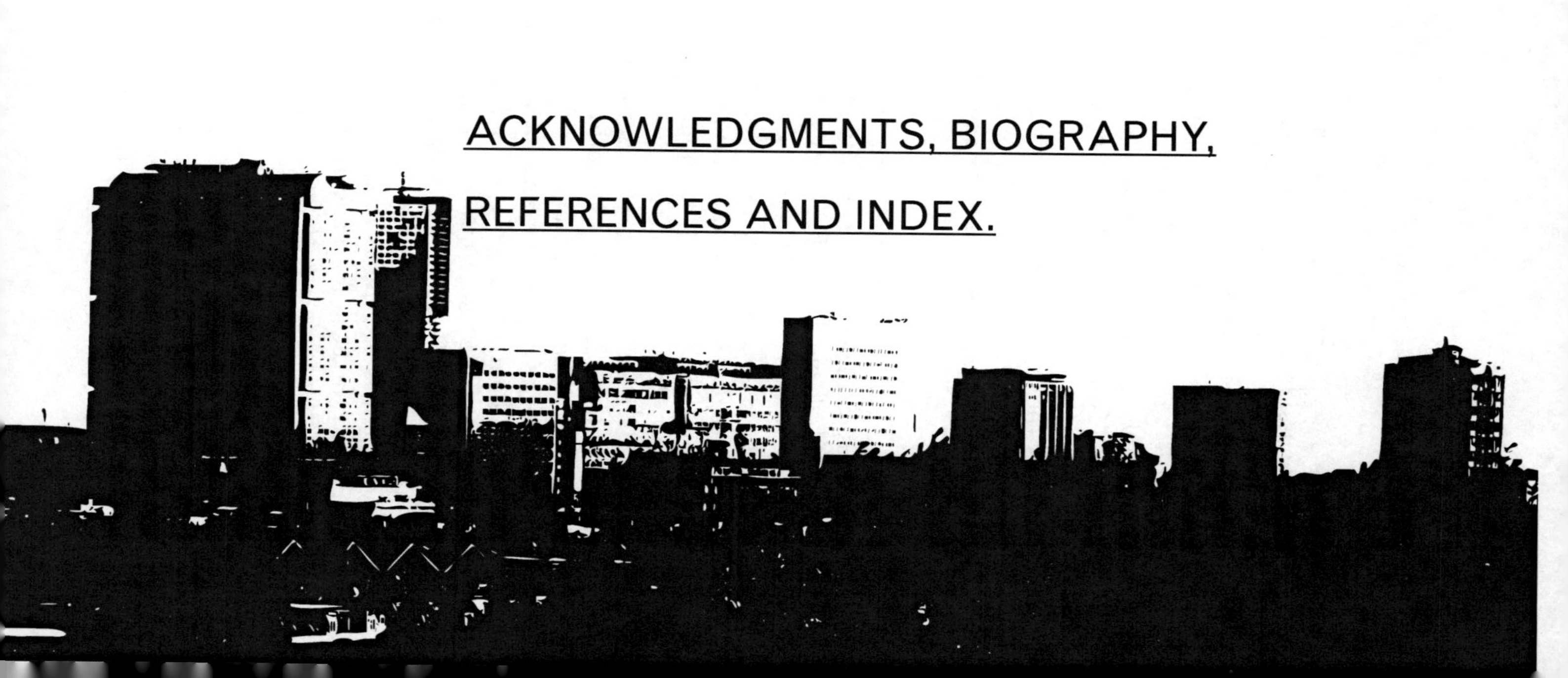

ACKNOWLEDGMENTS

Picture me, if you will, while I'm writing these words. I'm sitting outside my appartment in the glow of an afternoon sun in a beautiful Alpine village. The old lady from whom I rent this place is collecting grass from the pastures with a wooden hark. It is not to earn a living, she says, but just to help the farmers – so that they will help her when she needs them. It is a simple gesture that, over time, has created an intricate pattern of interactions that shapes life in this small place.

Multiply the factors that influence this image by a billion, and you can start to appreciate the extent of the endeavor facing those seeking to understand the social complexity that shapes our life.

I have a number of motives for writing this book. First, I am genuinely in awe of the complexity that makes up the reality of the world. To me, instead of being a dirty word, 'complexity' is inherent in real life and should be studied to the greatest level of detail possible. Second, my research experience tells me that quick fixes don't exist in the world of public decision-making. The intricacy of this field became clear to me as I thought about and researched complexity. I cannot pretend that a few simple guidelines or recommendations can serve as a magic wand to make things run smoothly but I can help analyzing the complexity decision-makers face every day.

Another thing that became clear to me when writing and teaching about the complexity sciences and public-decision making was the sheer number of publications on this topic produced each year in the fields of public administration, policy and political science, as well as the enormous variety of approaches and ideas that accompanied them. This made me realize the need for a succinct guide that comprehensively organized these ideas without being too simplistic, complicated or jargon-dense, and that also weeded out the slightly 'wilder' ones. In addition, I felt the need to relate ideas from the complexity sciences to that of the social sciences that have been around for a long time. Unlike some authors, I do not see the complexity sciences as presenting a revolution of thought. Rather, its main value for me lies in the fact that it provides a fairly coherent framework for integrating various ideas, including those that predate the complexity sciences, in a way that helps us gain a deeper understanding as to why public decision-making is such a complex subject. The key for this lies in how complexity deals with time and causation. As such, this book is a first but not final attempt to make sense of the complexity of public decision-making.

I believe that conjunction is often a more powerful explanation for how things come about than rational planning and purposeful action. This is very much the case for how this book came about. I wrote it myself, of course, but many people have contributed in one way or another to it and the ideas within. I would like to use this opportunity to thank them. First, I'm grateful to Kurt Richardson of Emergent Publications and the E:CO journal for believing in this book and giving me the opportunity to publish it. I also want to thank the Netherlands Organisation for Scientific Research (NWO) for awarding me a Veni-grant that allowed me to research the complexity of public decision-making both empirically and theoretically.

Many of the arguments in this book were shaped during intense discussions with Frank Boons, Peter Marks, Geert Teisman and Stefan Verweij and I thank them for their inspiration and insights. For working with me on several previous texts cited in this book, I'm appreciative of and indebted to the people mentioned above, as well as Arwin van Buuren, Nanny Bressers, Marcel van Gils, Joop Koppenjan, Rebecca Moody, Joris van der Voet and Danny Schipper. Frank, Peter, Danny, Wouter, Stefan and Göktug Morçöl all read parts of the book in a very detailed way and provided me with valuable feedback, for which I'm grateful - even when it meant I had to rewrite entire chapters that I had already crossed off my list of things to do. In addition, my thoughts and ideas were improved (or rejected, in some cases!) through collaboration and discussions with both scientists and practitioners, in particular: David Byrne, Jean-Marie Buijs, Thomas Catlaw, Ytsen Deelstra, Jurian Edelenbos, Jasper Eshuis, Yushim Kim, Erik-Hans Klijn, Chris Koliba, Jack Meek, Göktug Morçöl, Bonno Pel, Wart Rauws, Gert de Roo, Geert Teisman, and Vasco Vasic. I want to especially thank Zeno van den Broek who unleashed all of his qualities as an all-round artist for the design and artwork of this book.

It is entirely unfair that family and friends are, in the greatest scientific tradition, placed last in the list of people to thank. I want to thank my parents for their unconditional support in whatever I have chosen to do and my friends for bearing with me while I was doing it. A very special thanks goes to my wife Kwang for walking with me across the mountains and through the valleys of life; and to Trihn, who came to us when this book was not quite yet done and who reminded me that neither planning nor good intent can withstand the real dynamics of life. I dedicate this book to the both of you!

Rotterdam & Saas Fee,

LG

BIOGRAPHY
LASSE GERRITS

REFERENCES

Alchian, A. A. (1950). Uncertainty, evolution, and economic theory. *The Journal of Political Economy, 58*(3), 211-221.

Aldrich, H. E., & Ruef, M. (1999). *Organizations evolving.* London: Sage.

Algemene Rekenkamer. (2003). *Rapport risicoreservering HSL-Zuid en Betuweroute* No. 28 724. Den Haag: Algemene Rekenkamer.

Algemene-Rekenkamer. (2000). *Verdieping Westerschelde.* Den Haag: Algemene Rekenkamer.

Allen, P. M. (1998). Evolving complexity in social science. In G. Altmann, & W. A. Koch (Eds.), *Systems: New paradigms for human sciences* (pp. 3-38). Berlin: Walter de Gruyter GmbH & Co.

Allen, P., & Strathern, M. (2003). Evolution, emergence, and learning in complex systems. *Emergence, 5*(4), 8-33.

Arthur, W. B. (1994). *Increasing returns and path dependence in the economy.* Ann Arbor: The University of Michigan Press.

Arthur, W. B., & Durlauf, S. N. (Eds.). (1997). *The economy as an evolving complex system II.* Reading: Addison-Wesley.

Ashmos, D. P., Duchon, D., & McDaniel, R. R. (2000). Organizational responses to complexity: The effect on organizational performance. *Journal of Organizational Change, 13*(6), 577-594.

Aspa, G. (2006). Portraying, classifying and understanding the emerging landscapes in the post-industrial city. *Cities, 23*(5), 311-330.

Atkinson, R. (2004). The evidence on the impact of gentrification: New lessons for the urban renaissance? *European Journal of Housing Policy, 4*(1), 107-131.

Avelino, F. (2011). Power in transition: An interdisciplinary framework to study power in relation to structural change. (PhD, Erasmus University Rotterdam).

Axelrod, R. (1986). An evolutionary approach to norms. *The American Political Science Review,* 1095-1111.

Axelrod, R. M. (1997). *The complexity of cooperation: Agent-based models of competition and collaboration.* Princeton: Princeton University Press.

Axelrod, R. (1984). *The evolution of cooperation.* New York: BasicBook.

Ayal, S., & Zakay, D. (2009). The perceived diversity heuristic: The case of pseudodiversity. *Journal of Personality and Social Psychology, 96*(3), 559.

Ayres, R. U. (2004). On the life cycle metaphor: Where ecology and economics diverge. *Ecological Economics, 48*(1), 425-438.

Barnaby, M. (2002). Heuristics as social tools. *New Ideas in Psychology, 20*(1), 49-57.

Barton, J., Emery, M., Flood, R. L., Selsky, J. W., & Wolstenholme, E. (2004). A maturing of systems thinking? Evidence from three perspectives. *Systemic Practice and Action Research, 17*(1), 3-36.

Batty, M. (2010). Complexity in city systems: Understanding, evolution and design. In G. De Roo, & E. A. Silva (Eds.), *A planner's encounter with complexity* (pp. 99-122). Surrey: Ashgate.

Baumgartner, F. R., & Jones, B. D. (1993). *Agendas and instability in American politics*. Chicago: The University of Chicago Press.

Beahrs, J. O. (1992). Paradoxical effects in political systems. *Political Psychology, 13*(4), pp. 755-769.

Bednarz, J. (1984). Complexity and intersubjectivity: Towards the theory of Niklas Luhmann. *Human Studies, 7*(1-4), 55-69.

Beer, S. (1981). *Brain of the firm*. Chichester: John Wiley & Sons Ltd.

Beer, S. (1985) *Diagnosing the system for organizations*. Chichester: John Wiley and Sons Ltd.

Befani, B., Ledermann, S., & Sager, F. (2007). Realistic evaluation and QCA: Conceptual parallels and an empirical application. *Evaluation, 13*(2), 171-192.

Bendor, J. B., Kumar, S., & Siegel, D. A. (2009). Satisficing: A 'pretty good' heuristic. *The BE Journal of Theoretical Economics, 9*(1), 1-36.

Bergh, E. v. d., Damme, S. v., Graveland, J., Jong, D. d., Baten, I., & Meire, P. (2005). Ecological rehabilitation of the Schelde estuary (the Netherlands–Belgium; northwest Europe): Linking ecology, safety against floods, and accessibility for port development. *Restoration Ecology, 13*(1), 204-214.

Bergh, J. C. J. M. v. d. (2000). Ecological economics: Themes, approaches and differences with environmental economics. *Regional Environmental Change, 2*(1), 13-23.

Bergh, J. C. J. M. v. d. (2004). Evolutionary thinking in environmental economics: Retrospect and prospect. In J. Foster, & W. Hölzl (Eds.), *Applied evolutionary economics and complex systems* (pp. 293). Cheltenham: Edward Elgar.

Bergh, J. C. J. M. v. d., Faber, A., Idenburg, A. M., & Oosterhuis, F. H. (2005). *Survival of the greenest* No. 550006002/2005. Bilthoven: Milieu- en Natuurplanbureau.

Bergh, J. C. J. M. v. d., & Kallis, G. (2009). *Evolutionary policy*. Max Planck Insitute of Economics.

Bergh, J. C. J. M. v. d., & Gowdy, J. M. (2000). Evolutionary theories in environmental and resource economics: Approaches and applications. *Environmental and Resource Economics, 17*, 37-57.

Berlyne, D. E. (1971). *Aesthetics and psychobiology*. Appleton-Century-Crofts.

Bertolini, L. (2010). Complex systems, evolutionary planning? In G. d. Roo, & E. A. Silva (Eds.), *A planner's encounter with complexity* (pp. 81-98). Farnham: Ashgate.

Bhaskar, R. (1979). *The possibilities of naturalism*. Atlantic Highlands, NJ: Humanities Press.

Bhaskar, R. (1988). *The possibility of naturalism*. London: Routledge.

Bhaskar, R. (2008). *A realist theory of science*. New York: Routledge.

Bingham, R. D., & Kimble, D. (1995). The industrial composition of edge cities and downtowns: The new urban reality. *Economic Development Quarterly, 9*(3), 259-272.

Blackman, T. (2006). *Placing health: Neighbourhood renewal, health improvement and complexity*. Bristol: The Policy Press.

Blatter, J., & Blume, T. (2008). In search of co-variance, causal mechanisms or congruence? towards a plural understanding of case studies. *Swiss Political Science Review, 14*(2), 315-356.

Blatter, J. (2003). Beyond hierarchies and networks: Institutional logics and change in transboundary spaces. *Governance, 16*(4), 503-526.

Blom, T., & Haas, B. (1997). De ondraaglijke lichtheid van systemen. over de grondslagen van het Luhmanniaanse denken. *Tijdschrift Voor Sociologie, 17*(2), 187-204.

Boisot, M. (2000). Is there a complexity beyond the reach of strategy? *Emergence, 2*(1), 114-134.

Bontje, M., & Burdack, J. (2005). Edge cities, European-style: Examples from Paris and the Randstad. *Cities, 22*(4), 317-330.

Branda, N. (2008). Response to paper "Systems thinking" by D. cabrera et al.: Conceptualizing systems thinking in evaluation. *Evaluation and Program Planning, 31*(3), 329-331.

Brans, M., & Rossbach, S. (1997). The autopoiesis of administrative systems: Niklas Luhmann on public administration and public policy. *Public Administration, 75*(3), 417-439.

Briguglio, L., Cordina, G., Farrugia, N., & Vella, S. (2009). Economic vulnerability and resilience: Concepts and measurements. *Oxford Development Studies, 37*(3), 229-247.

Bronstein, P. M. (2000). Ecological heuristics for learning. *Behavioral and Brain Sciences, 23*(02), 251.

Buijs, M. J., Eshuis, J., & Byrne, D. S. (2009). Approaches to researching complexity in public management. In G. R. Teisman, M. W. Van

Buuren & L. M. Gerrits (Eds.), *Managing complex governance systems: Dynamics, self-organization and coevolution in public investments* (pp. 37-55). New York: Routledge.

Burgers, J. (2002). De gefragmenteerde stad. *Amsterdams Sociologisch Tijdschrift, 28*(4)

Burns, D. (2007). *Systemic action research: A strategy for whole system change*. Bristol: Policy Press.

Byrne, D. S. (1998). *Complexity theory and the social sciences: An introduction*. London: Routledge.

Byrne, D. S. (2001). *Understanding the urban*. Palgrave: Inglaterra.

Byrne, D. S. (2004). Complexity theory and social research. *Social Research Update*

Byrne, D. S. (2005). Complexity, configurations and cases. *Theory, Culture & Society, 22*(5), 95-111.

Byrne, D. S. (2011a). *Applying social science: The role of social research in politics, policy and practice*. Bristol: The Policy Press.

Byrne, D. S. (2011b). Exploring organizational effectiveness: The value of complex realism as a frame of reference and systematic comparison as a method. In P. Allen, S. Maguire & B. McKelvey (Eds.), *The sage handbook of complexity and management* (pp. 131-141). London: SAGE Publications Ltd.

Byrne, D. S, & Ragin, C. C. (2009). *The Sage handbook of case-based methods*. London: Sage Publications Ltd.

Cabrera, D., & Colosi, L. (2008). Distinctions, systems, relationships, and perspectives (DSRP): A theory of thinking and of things. *Evaluation and Program Planning, 31*(3), 311-317.

Cabrera, D., Colosi, L., & Lobdell, C. (2008). Systems thinking. *Evaluation and Program Planning, 31*(3), 299-310.

Callaghan, G. (2008). Evaluation and negotiated order. *Evaluation, 14*(4), 399-411.

Cameron, S., & Doling, J. (1994). Housing neighbourhoods and urban regeneration. *Urban Studies, 31*(7), 1211-1223.

Cantor, G. N. (1982). Review: The possibility of naturalism: A philosophical critique of the contemporary human sciences. *The Philosophical Quarterly, 32*(128), pp. 280-281.

Certomà, C. (2006). Ecology, environmentalism and system theory. *Kybernetes, 35*(6), 915-921.

Checkland, P., & Holwell, S. (1998). *Information, systems and information systems - making sense of the field*. Hoboken: Wiley.

Checkland, P. (1981). *Systems thinking, systems practice*. Chichester etc.: John Wiley & Sons.

Chettiparamb, A. (2006). Metaphors in complexity theory and

planning. *Planning Theory, 5*(1), 71-91.
Chong, D., Citrin, J., & Conley, P. (2001). When self-interest matters. *Political Psychology, 22*(3), 541-570.
Cilliers, P. (1998). *Complexity and postmodernism. understanding complex systems*. London: Routledge.

Cilliers, P. (2001). Boundaries, hierarchies and networks in complex systems. *International Journal of Innovation Management, 5*, 135-148.
Cilliers, P. (2002). Why we cannot know complex things completely. *Emergence, 4*(1/2), 77-84.
Cilliers, P. (2005a). Complexity, deconstruction and relativism. *Theory, Culture and Society, 22*(5), 255-267
Cilliers, P. (2005b). Knowledge, limits and boundaries. *Futures, 37*(7), 605-613.
Clarke, D. D., & Crossland, J. (1985). *Action systems: An introduction to the analysis of complex behaviour*. London: Methuen.
Cobb, R. W., & Elder, C. D. (1972). *Participation in American politics: The dynamics of agenda-building*. Baltimore: Johns Hopkins University Press.
Cohen, M. D., March, J. G., & Olsen, J. P. (1972). A garbage can model of organizational choice. *Administrative Science Quarterly*, , 1-25.
Conner, D. R. (1998). *Leading at the edge of chaos: How to create the nimble organization*. London: Wiley.
Cooksey, R. W. (2000). Mapping the texture of managerial decision making: A complex dynamic decision perspective. *Emergence, 2*(2), 102-122.
Cooper, R. (2000). Simple heuristics could make us smart; but which heuristics do we apply when? *Behavioral and Brain Sciences, 23*(05), 746.
Corning, P. A. (1983). *The synergism hypothesis: A theory of progressive evolution*. New York: McGraw-Hill.
Cox, R., Wadsworth, R., & Thomson, A. (2003). Long-term changes in salt marsh extent affected by channel deepening in a modified estuary. *Continental Shelf Research, 23*(17-19), 1833-1846.
Crozier, M. (1964). *The bureaucratic phenomenon*. Chicago: The University of Chicago Press.
Cyert, R. M., & James, J. G. (1992). *A behavioral theory of the firm (2nd ed.)*. Oxford: Blackwell Publishers Ltd.
Dahl, R. A. (1958). A critique of the ruling elite model. *The American Political Science Review, 52*(2), 463-469.
Dahl, R. A. (1961). The behavioral approach in political science: Epitaph for a monument to a successful protest. *The American*

Political Science Review, 55(4), 763-772.

Damasio A. R., Tranel D., Damasio H. (1990) Individuals with sociopathic behavior caused by frontal damage fail to respond autonomically to social stimuli. *Behav Brain Res* 41:81–94.

Daneke, G. A. (1990). A science of public administration?. *Public Administration Review, 50*(3), 383-392.

Darwin, C. (1859 / 1985). *The origin of species by means of natural selection; or the preservation of favoured races in the struggle for life*. Essex: Penguin Books Ltd.

David, P. A. (1985). Clio and the economics of QWERTY. *The American Economic Review, 75*(2), 332-337.

De Meur, G., Rihoux, B., & Yamasaki, S. (2009). Addressing the critiques of QCA. In B. Rihoux, & C. C. Ragin (Eds.), *Configurational comparative methods: qualitative comparative analysis (QCA) and related techniques* (pp. 147-166). London: Sage.

De Roo, G. (2010). Being or becoming? that is the question! confronting complexity with contemporary planning theory. In G. De Roo, & E. A. Silva (Eds.), *A planner's encounter with complexity* (pp. 19-40). Surrey: Ashgate.

De Roo, G., Hillier, J., & Van Wezemael, J. (Eds.). (2012). *Planning & complexity: Systems, assemblages and simulations*. Farnham: Ashgate.

De Roo, G., & Schwartz, M. (Eds.). (2001). *Omgevingsplanning, een innovatief proces: Over integratie, participatie, omgevingsplannen en de gebiedsgerichte aanpak*. Den Haag: Sdu.

De Rynck, F. (1994). *Streekontwikkeling in Vlaanderen: Bestuurskundig bekeken: besturingsverhoudingen en beleidsnetwerken in bovenlokale ruimtes met illustratief case-onderzoek in het zuiden van de provincie West-Vlaanderen*. Leuven: Katholieke Universiteit Leuven.

Dechert, C. A. (1966). Positive feedback in political and international systems. *American Behavioral Scientist, 9*(8), 8-14.

Desouza, K. C., & Lin, Y. (2011). Towards evidence-driven policy design: Complex adaptive systems and computational modeling. *The Innovation Journal: The Public Sector Innovation Journal, 16*(1)

Diamond, J. (2005). *Collapse: How societies choose to fail or succeed*. New York: Viking Press.

Diamond, M. A. (1985). The social character of bureaucracy: Anxiety and ritualistic defense. *Political Psychology, 6*(4), pp. 663-679.

Diehl, E., & Sterman, J. D. (1995). Effects of feedback complexity on dynamic decision-making. *Organizational Behavior and Human Decision Processes, 62*(2), 198-215.

Doak, J., & Karadimitriou, N. (2007). (Re) development, complexity and networks: A framework for research. *Urban Studies, 44*(2), 209.

Dopfer, K. (2005). *The evolutionary foundations of economics.* Cambridge: Cambridge University Press.

Dornish, D. (2002). The evolution of post-socialist projects: Trajectory shift and transitional capacity in a Polish region. *Regional Studies, 36*(3), 307-321.

Dyson, S. B., & Preston, T. (2006). Individual characteristics of political leaders and the use of analogy in foreign policy decision-making. *Political Psychology, 27*(2), 265-288.

Earl, P. E., & Wakeley, T. (2010). Alternative perspectives on connections in economic systems. *Journal of Evolutionary Economics, 20*(2), 163-183.

Easton, G. (2010). Critical realism in case study research. *Industrial Marketing Management, 39*(1), 118-128.

Edwards, A., & Schaap, L. (2006). *Burgerparticipatie in Rotterdam.* Rotterdam: Erasmus Universiteit Rotterdam.

Ehin, C. (2010; 2010). Muddling through engaging our innate heuristics. *The Journal for Quality and Participation, 33*(3), 32-32-36.

Elder-Vass, D. (2005). Emergence and the realist account of cause. *Journal of Critical Realism, 4*(2), 315-338.

Eldredge, N., & Gould, S. J. (1972). Punctuated equilibria: An alternative to phyletic gradualism. In T. J. M. Schopf (Ed.), *Models in paleobiology* (pp. 82-115). San Fransisco: Cooper & Co.

Ellen, G., Gerrits, L., & Slob, A. (2007). Risk perception and risk communication. In S. Heise (Ed.), *Sediment risk management and communication* (pp. 292). Amsterdam: Elsevier.

Elster, J. (1976). A note on hysteresis in the social sciences. *Synthese, 33*(1), 371-391.

Elster, J. (2007). *Explaining social behavior: More nuts and bolts for the social sciences.* Cambridge: Cambridge University Press.

Epley, N., & Gilovich, T. (2006). The anchoring-and-adjustment heuristic. *Psychological Science (Wiley-Blackwell), 17*(4), 311-318.

Epstein, S. (1994). Integration of the cognitive and the psychodynamic unconscious. *American Psychologist, 49*(8), 709.

Erdfelder, E., & Brandt, M. (2000). How good are fast and frugal inference heuristics in case of limited knowledge? *Behavioral and Brain Sciences, 23*(05), 747.

Eshuis, J., & Edelenbos, J. (2009). Branding in urban regeneration. *Journal of Urban Regeneration and Renewal, 2*(3), 272-282.

Farnham, B. (1990). Political cognition and decision-making. *Political Psychology, 11*(1), pp. 83-111.

Fath, B. D., & Grant, W. E. (2007). Ecosystems as evolutionary complex systems: Network analysis of fitness models. *Environmental Modelling & Software, 22*(5), 693-700.

Fischer, F. (1998). Beyond empiricism: Policy inquiry in postpositivist tradtion. *Policy Studies Journal, 26*(1)

Fischer, F. (2003). *Reframing public policy: Discursive politics and deliberative practices*. Oxford: Oxford University Press.

Fischer, F., & Forester, J. (Eds.). (1987). *Confronting values in policy analysis: The politics of criteria*. London: Sage.

Fischer, F., & Forester, J. (Eds.). (1993). *The argumentative turn in policy analysis and planning*. Durham: Duke University Press.

Fischhoff, B. (1975). Hindsight is not equal to foresight: The effect of outcome knowledge on judgment under uncertainty. *Journal of Experimental Psychology: Human Perception and Performance, 1*(3), 288.

Flood, R. L. (1999a). *Rethinking the fifth discipline: Learning within the unknowable*. London: Routledge.

Flood, R. L. (1999b). Knowing of the unknowable. *Systemic Practice and Action Research, 12*(3), 247-256.

Flood, R. L., & Jackson, M. C. (1991). *Creative problem solving: Total systems intervention*. Chichester: John Wiley & Sons.

Flyvbjerg, B. (2002). *Making social science matter: Why social inquiry fails and how it can succeed again*. Cambridge: Cambridge University Press.

Flyvbjerg, B., Bruzelius, N., & Rothengatter, W. (2006). *Megaprojects and risk. an anatomy of ambition*. Cambridge: Cambridge University Press.

Flyvbjerg, B., Skamris Holm, M. K., & Buhl, S. L. (2005). How (in) accurate are demand forecasts in public work projects? the case of transportation. *Journal of the American Planning Association, 71*(2), 131-146.

Folke, C. (2006). Resilience: The emergence of a perspective for social-ecological systems analysis. *Global Environmental Change, 16*(1), 253-267.

Folke, C., Hahn, T., Olsson, P., & Norberg, J. (2005). Adaptive governance of social-ecological systems. *Annual Review of Environment and Resources, 30*(1), 441-473.

Forrester, J. W. (1971). *World dynamics*. Cambridge, MA: Wright-

Allen Press.

Foster, J. (2005). From simplistic to complex systems in economics. *Cambridge Journal of Economics, 29*(6), 873-892.

Foster, J., & Hölzl, W. (Eds.). (2004). *Applied evolutionary economics and complex systems*. Cheltenham UK, Northampton USA: Edward Elgar Publishing.

François, C. (1999). Systemics and cybernetics in a historical perspective. *Systems Research and Behavioral Science, 16*(3), 203-219.

Frank, H., & Ganapati, S. (2008). Good intentions, unintended consequences: Impact of Adker consent decree on Miami-Dade county's subsidized housing. *Urban Affairs Review, 44*(1), 57-84.

Frenken, K. (2006). A fitness landscape approach to technological complexity, modularity, and vertical disintegration. *Structural Change and Economic Dynamics 17*, 288-305.

Frenken, K., & Nuvolari, A. (2004). Entropy statistics as a framework to analyse technological evolution. In J. Foster, & W. Hölzl (Eds.), *Applied evolutionary economics and complex systems* (pp. 95-132). Cheltenham, UK and Northampton, MA: Edward Elgar Publishing.

Fuchs, C. (2002). Concepts of social self-organisation. *INTAS Project 'Human Strategies in Complexity'-Report. HSIC Paper,* (4)

Garreau, J. (1991). *Edge city: Life on the new frontier*. New York: Anchor Books.

Geddes, P. (1915). *Cities in evolution: An introduction to the town planning movement and to the study of civics*. London: Williams & Norgate.

Gell-Mann, M. (1995). What is complexity? *Complexity, 1*(1), 16-19.

Gell-Mann, M. (1996). Let's call it plectics. *Complexity, 1*(5)

Geoff, E. (2010). Critical realism in case study research. *Industrial Marketing Management, 39*(1), 118-128.

George, A. L., & Bennett, A. (2005). *Case studies and theory development in the social sciences* The MIT Press.

Gerald, M. (2008). Response to paper "Systems thinking" by D. Cabrera et al.: The unification of systems thinking: Is there gold at the end of the rainbow? *Evaluation and Program Planning, 31*(3), 317-321.

Gerrits, L. M. (2008). *The gentle art of coevolution: A complexity theory perspective on decision making over estuaries in Germany, Belgium and the Netherlands*. Rotterdam: Erasmus University Rotterdam.

Gerrits, L. M. (2010). Public decision-making as coevolution.

Emergence, 12(1), 19-28.

Gerrits, L. M. (2011). A coevolutionary revision of decision making processes: An analysis of port extensions in Germany, Belgium and the Netherlands. *Public Administration Quarterly, 35*(3), 309-339.

Gerrits, L., & Marks, P. M. (2008). Complex bounded rationality in dyke construction; path-dependency and lock-in in the emergence of the geometry of the Zeeland delta. *Land Use Policy, 25*(3), 330-337.

Gerrits, L. M., & Koppenjan, J. F. M. (2011, Aanbesteden hoofdspoor biedt kansen in de regio. *OV Magazine, 5*

Gerrits, L. M., & Verweij, S. (2012). Critical realism as a meta-framework for understanding the relationships between complexity and qualitative comparative analysis. *Journal of Critical Realism (forthc.)*

Gerrits, L. M., & Meek, J. W. (2012). Propositions for complexity science. In L. M. Gerrits, P. Marks & F. A. A. Boons (Eds.), *Public administration in complexity*. Litchfield Park, AZ: Emergent Publications.

Gersich, C. J. G. (1991). Revolutionary change theories: A multilevel exploration of the punctuated equilibrium paradigm. *Academy of Management Review, 16*(1), 10-36.

Giddens, A. (1984). *The constitution of society: Outline of the theory of structuration* University of California Press.

Gigerenzer, G., & Gaissmaier, W. (2011). Heuristic decision making. *Annual Review of Psychology, 62*, 451-482.

Gigerenzer, G., & Goldstein, D. G. (1996). Reasoning the fast and frugal way: Models of bounded rationality. *Psychological Review, 103*(4), 650.

Gill, T. (2008). Reflections on researching the rugged fitness landscape. *Informing Science, 11*, 165.

Gleick, J. (1987). *Chaos*. London: Vintage.

Goldstein, J. (1999). Emergence as a construct: History and issues. *Emergence, 1*(1), 49-72.

Gowda, M. V. R. (1999; 1999). Heuristics, biases, and the regulation of risk. *Policy Sciences, 32*(1), 59-59-78.

Graham, R., & Seltzer, J. (1979). An application of catastrophe theory to management science process. *Omega, 7*(1), 61-66.

Greener, I. (2002). *Theorising path dependence: How does history come to matter in organisations, and what can we do about it?* Unpublished manuscript.

Grofman, B., & Schneider, C. Q. (2009). An introduction to crisp set QCA with a comparison to binary logistic regression. *Political*

Research Quarterly, 62(4), 662-672.

Gual, M. A., & Norgaard, R. B. (2010). Bridging ecological and social systems coevolution: A review and proposal. *Ecological Economics, 69*(4), 707–717.

Gulick, L. (1984). The metaphors of public administration. *Public Administration Quarterly, 8*(3), 369-381.

Gunderson, L. H. (2001). Managing surprising ecosystems in southern Florida. *Ecological Economics, 37*(3), 333-484.

Haan, F. J. d. (2010). *Towards transition theory.* Rotterdam: Erasmus University Rotterdam.

Hahn, R. W., & Tetlock, P. C. (2005). Using information markets to improve public decision making. *Harvard Journal of Law & Public Policy, 29*(1), 213-289.

Hahn, U., Frost, J., & Maio, G. (2005). What's in a heuristic? *Behavioral and Brain Sciences, 28*(04), 551.

Hajer, M. A., & Wagenaar, H. (2003). *Deliberative policy analysis: Understanding governance in the network society.* Cambridge: Cambridge University Press.

Hall, P. A. (1993). Policy paradigms, social learning, and the state: The case of economic policymaking in Britain. *Comparative Politics, 25*(3), 275-296.

Hamdi, N. (2004). *Small change. about the art of practice and the limits of planning in cities.* London: Earthscan.

Hammersley, M. (2008). Causality as conundrum: The case of qualitative inquiry. *Methodological Innovations Online, 2*(3) Retrieved from http://erdt.plymouth.ac.uk/mionline/public_html/viewarticle.php?id=63&layout=html

Hammerstein, P. (2000). Evolution, learning, games, and simple heuristics. *Behavioral and Brain Sciences, 23*(05), 752.

Hammond, R. A., & Axelrod, R. (2006). The evolution of ethnocentrism. *Journal of Conflict Resolution, 50*(6), 926-936.

Harre, R. (1976). [Untitled]. *Mind, 85*(340), pp. 627-630.

Harter, N. (2006). Leadership as the promise of simplification. *Emergence: Complexity & Organization, 8*(4), 77-87.

Hartigh, E. (2005). Increasing returns and firm performance: An empirical study. (PhD, Erasmus University Rotterdam).

Hartvigsen, G., Kinzig, A., & Peterson, G. (1998). Use and analysis of complex adaptive systems in ecosystem science: Overview of special section. *Ecosystems, 1*(5), 427-430.

Haynes, P. (2001). Complexity, quantification and the management of policy. *Social Issues, 1*(2)

Haynes, P. (2003). *Managing complexity in the public services.*

Maidenhead: Open University Press.
Hermann, M. G. (1979). Indicators of stress in policymakers during foreign policy crises. *Political Psychology, 1*(1), pp. 27-46.
Hertogh, M. J. C. M., & Westerveld, E. (2010). *Playing with complexity: Management and organisation of large infrastructure projects.* Rotterdam: Erasmus University.
Hilbig, B. E., & Pohl, R. F. (2008). Recognizing users of the recognition heuristic. *Experimental Psychology (Formerly Zeitschrift Für Experimentelle Psychologie), 55*(6), 394-401.
Hinssen, J. J. P. (1994). Verstarring en dynamiek van de bureaucratie. *Bestuurskunde, 3*(8)
Hird, M. J. (2010). Coevolution, symbiosis and sociology. *Ecological Economics, 69*(4), 737-742.
Hodgson, G. M. (2010). Darwinian coevolution of organizations and the environment. *Ecological Economics, 69*(4), 700-706.
Holland, J. H. (1995). *Hidden order: How adaptation builds complexity.* Jackson: Perseus Books.
Holland, J. H. (2006). Studying complex adaptive systems. *Journal of Systems Science and Complexity, 19*(1), 1-8.
Holling, C. S. (1973). Resilience and stability of ecological systems. *Annual Review of Ecology and Systematics, 4*, 1-23.
Hoogerwerf, A. (1984). Beleid berust op veronderstellingen: De beleidstheorie. *Acta Politica, 14*(4), 493-531.
Houghton, L. (2009). Generalization and systemic epistemology: Why should it make sense? *Systems Research and Behavioral Science, 26*(1), 99-108.
Hrebiniak, L. G., & Joyce, W. F. (1985). Organizational adaptation: Strategic choice and environmental determinism. *Administrative Science Quarterly, 30*(3), 336-348.
Huber, O. (2000). What's in the adaptive toolbox: Global heuristics or more elementary components? *Behavioral and Brain Sciences, 23*(05), 755.
Isaac, A. G. (1994). Hysteresis. In P. Arestis, & M. Sawyer (Eds.), *The Elgar companion to radical political economy.* London: Edward Elgar.
Jay, F. (2008). A response to paper "Systems thinking" by D. Cabrera et al.: Additional thoughts on systems thinking. *Evaluation and Program Planning, 31*(3), 333-334.
Jervis, R. (1992). Political implications of loss aversion. *Political Psychology, 13*(2, Special Issue: Prospect Theory and Political Psychology), pp. 187-204.
Jespersen-Groth, J., Potthoff, D., Clausen, J., Huisman, D., Kroon, L.,

Maróti, G., et al. (2009). Disruption management in passenger railway transportation. In R. K. Ahuja, R. H. Moehring & C. Zaroliagis (Eds.), *Robust and online large-scale optimization* (pp. 399-421). Berlin: Springer.

John, P. (1999). Ideas and interests; agendas and implementation: An evolutionary explanation of policy change in british local government finance. *The British Journal of Politics & International Relations, 1*(1), 39-62.

Johnston, E., Kim, Y., & Ayyangar, M. (2007). Intending the unintended: The act of building agent-based models as a regular source of knowledge generation. *Interdisciplinary Description of Complex Systems, 5*(2), 81-91.

Kahneman, D., & Frederick, S. (2002). Representativeness revisited: Attribute substitution in intuitive judgment. *Heuristics and Biases: The Psychology of Intuitive Judgment,* , 49-81.

Kahneman, D., & Tversky, A. (1972). Subjective probability: A judgment of representativeness. *Cognitive Psychology, 3*(3), 430-454.

Kahneman, D., & Tversky, A. (1979). Prospect theory: An analysis of decision under risk. *Econometrica: Journal of the Econometric Society,* , 263-291.

Kallis, G., & Norgaard, R. B. (2010). Coevolutionary ecological economics. *Ecological Economics, 69*(4), 690-699.

Kauffman, S. A., & Levin, S. (1987). Towards a general theory of adaptive walks on rugged landscapes. *Journal of Theoretical Biology, 128*(1), 11-45.

Kauffman, S. A., & Johnsen, S. (1991). Coevolution to the edge of chaos: Coupled fitness landscapes, poised states, and coevolutionary avalanches. *Journal of Theoretical Biology, 149*(4), 467-505.

Kauffman, S. A. (1993). *The origins of order.* New York, Oxford: Oxford University Press.

Kaufman, H. (1991). *Time, chance, and organizations.* New Jersey: Chatham House.

Kerr, P. (2002). Saved from extinction: Evolutionary theorising, politics and the state. *British Journal of Politics and International Relations, 4*(2), 330-358.

Kickert, W. J. M., Klijn, E. H., & Koppenjan, J. F. M. (Eds.). (1997). *Managing complex networks: Strategies for the public sector.* London: Sage Publications Ltd.

Kickert, W. J. M. (1991). *Complexiteit, zelfsturing en dynamiek.* Rotterdam: Erasmus Universiteit Rotterdam.

Kiel, L. D. (1989). Nonequilibrium theory and its implications for public administration. *Public Administration Review, 49*(6), 544-

551.
Kiel, L. D., & Elliott, E. (Eds.). (2005). *Chaos theory in the social sciences.* Ann Arbor: University of Michigan Press.
Kingdon, J. W. (1984). Agendas, alternatives, and public policies. Little, Brown, and Company, 240.
Kirman, A. (2004). The structure of economic interaction: Individual and collective rationality. *Cognitive Economics: An Interdisciplinary Approach,* , 293.
Kiyonari, T., Tanida, S., & Yamagishi, T. (2000). Social exchange and reciprocity: Confusion or a heuristic? *Evolution and Human Behavior, 21*(6), 411-427.
Kjær, A. M. (2011). Rhode's contribution to governance theory: praise, criticism and the future governance debate. *Public Administration, 89*(1), 101-113.
Klein, R. J. T., Nicholls, R. J., & Thomalla, F. (2003). Resilience to natural hazards: How useful is this concept? *Global Environmental Change Part B: Environmental Hazards, 5*(1–2), 35-45.
Klein, R. J. T., Smit, M. J., Goosen, H., & Hulsbergen, C. H. (1998). Resilience and vulnerability: Coastal dynamics or Dutch dikes? *The Geographical Journal, 164*(3), 259-268.
Klijn, E. H. (1996). *Regels en sturing in netwerken: De invloed van netwerkregels op de herstructurering van naoorlogse wijken.* Delft: Eburon.
Klijn, E. H., & Snellen, I. (2009). Complexity theory and public administration. In G. R. Teisman, M. W. Van Buuren & L. M. Gerrits (Eds.), *Managing complex governance systems: Dynamics, self-organization and coevolution in public investments* (pp. 17-36). New York: Routledge.
Koehler, J. J., & Gershoff, A. D. (2003). Betrayal aversion: When agents of protection become agents of harm. *Organizational Behavior and Human Decision Processes, 90*(2), 244-261.
Koliba, C., Meek, J. W., & Zia, A. (2010). *Governance networks in public administration and public policy.* Boca Raton: CRC Press, Inc.
Koliba, C., & Zia, A. (2012). Theory testing using complex systems modeling in public administration and policy studies: Challenges and opportunities for a meta-theoretical research program. In L. M. Gerrits, P. M. Marks & F. A. A. Boons (Eds.), *Public administration in complexity.* Litchfield Park, AZ: Emergent Publications.
Koliba, C., Zia, A., & Lee, B. H. Y. (2011). Governance informatics: Managing the performance of inter-organizational governance networks. *The Innovation Journal: The Public Sector Innovation*

Journal, 16(1)

Koppenjan, J. F. M., & Klijn, E. H. (2004). *Managing uncertainties in networks*. London: Routledge.

Kotchen, M. J., & Young, O. R. (2007). Meeting the challenges of the anthropocene: Towards a science of coupled human-biophysical systems. *Global Environmental Change, 17*(2), 149-151.

Kruger, J., Wirtz, D., Van Boven, L., & Altermatt, T. W. (2004). The effort heuristic. *Journal of Experimental Social Psychology, 40*(1), 91-98.

Kruglanski, A. W. (1992). On methods of good judgment and good methods of judgment: Political decisions and the art of the possible. *Political Psychology, 13*(3), pp. 455-475.

Laermans, R. (1996). We kunnen ons geen alternatief voorstellen: Luhmanns analyse van de moderne maatschappij. *Tijdschrift Voor Sociologie, 2*, 145-161.

Lakoff, G., & Johnson, M. (2003). *Metaphors we live by*. Chicago University Press: Chicago.

Langton, C. G. (1986). Studying artificial life with cellular automata. *Physica D: Nonlinear Phenomena, 22*(1), 120-149.

Latour, B. (2005). *Reassembling the social: An introduction to actor-network-theory*. Oxford: Oxford University Press.

Lee, J., & Kim, J. (2007). Grounded theory analysis of e-government initiatives: Exploring perceptions of government authorities. *Government Information Quarterly, 24*(1), 135-147.

Leeuw, F. L. (2003). Reconstructing program theories: Methods available and problems to be solved. *The American Journal of Evaluation, 24*(1), 5-20.

Lefebure, E., & Letiche, H. (1999). Managing complexity from chaos: Uncertainty, knowledge and skills. *Emergence: Complexity & Organization, 1*(3), 7.

Levin, S. A. (1978). On the evolution of ecological parameters. In P. E. Brussard (Ed.), *Ecological genetics: The interface* (pp. 3-26). New York: Proceedings in Life Sciences, Springer.

Levin, S. A. (1998). Ecosystems and the biosphere as complex adaptive systems. *Ecosystems, 1*(5), 431-436.

Levin, S. A. (1999). *Fragile dominion. Complexity and the commons*. Reading, Massachusetts: Perseus Books.

Levinthal, D. A. (1997). Adaptation on rugged landscapes. *Management Science*, , 934-950.

Levinthal, D. A., & Warglien, M. (1999). Landscape design: Designing for local action in complex worlds. *Organization Science*, , 342-357.

Levitt, S. D., & List, J. A. (2007). What do laboratory experiments measuring social preferences reveal about the real world? *Journal of Economic Perspectives, 21*(2), 153-174.

Leydesdorff, L. (1997). The non-linear dynamics of sociological reflections. *International Sociology, 12*(1), 25-45.

Lindgren, K. (1997). Evolutionary dynamics in game-theoretic models. *Santa Fe Institute Studies in the Sciences of Complexity – Proceedings volume 27.* pp. 337-368.

Loewenstein, G. F., Weber, E. U., Hsee, C. K., & Welch, N. (2001). Risk as feelings. *Psychological Bulletin, 127*(2), 267.

Losch, A. (2009). On the origins of critical realism. *Theology and Science, 7*(1), 85-106.

Luce, R. D. (2000). Fast, frugal, and surprisingly accurate heuristics. *Behavioral and Brain Sciences, 23*(05), 757.

Luhmann, N. (1970). *Soziologische aufklärung: Aufsätze zur Theorie sozialer Systeme.* Köln/Opladen: Westdeutscher Verlag.

Luhmann, N. (1977). Differentiation of society. *Canadian Journal of Sociology/Cahiers Canadiens De Sociologie,* , 29-53.

Luhmann, N. (1981). *Soziologische Aufklärung 3. Soziales System, Gesellschaft, Organisation* . Opladen: Westdeutscher Verlag.

Luhmann, N. (1984). *Soziale systeme: Grundriß einer allgemeinen theorie.* Frankfurt am Main: Suhrkamp.

Luhmann, N. (1995). *Social systems* [Soziale systeme: Grundriß einer allgemeinen Theorie] (J. Bednarz, D. Baecker Trans.). Stanford CA: Stanford University Press.

Lukes, S. (1974). *Power: A radical view.* Basingstoke: Palgrave Macmillan.

Maasland, E., Koppenjan, J. F. M., & Gerrits, L. M. (2011). *Het nieuwe spoorplan: Inschatting besparingen en analyse organisatorische en bestuurlijke consequenties.* Rotterdam: SEOR Erasmus School of Economics.

Mackenzie, A. (2005). The problem of the attractor: A singular generality between sciences and social theory. *Theory, Culture & Society, 22*(45), 45-65.

Mackie, J. L. (1980). *The cement of the universe: A study of causation.* Oxford: Oxford University Press.

Maguire, S., & McKelvey, B. (1999). Complexity and management: Moving from fad to firm foundations. *Emergence, 1*(2), 19-61.

Majone, G., & Quade, E. S. (Eds.). (1980). *Pitfalls of analysis.* Chichester: Wiley.

Manner, M., & Gowdy, J. (2010). The evolution of social and moral behavior: Evolutionary insights for public policy. *Ecological*

Economics, 69(4), 753-761.

March, J. G. (1991). Exploration and exploitation in organizational learning. *Organization Science, 2*(1), 71-87.

March, J. G. (1994). *A primer on decision making: How decisions happen*. New York: The free press.

Marcus, G. E., MacKuen, M., & Neuman, W. R. (2011). Parsimony and complexity: Developing and testing theories of affective intelligence. *Political Psychology, 32*(2), 323-336.

Maring, L., Gerrits, L., & Joziasse, J. (2005). *Elbe river basin characterisation. AquaTerra integrator I 1.1d* No. 505428). Apeldoorn: TNO Environment and Geosciences.

Marion, R. (1999). *The edge of organization*. Thousand Oaks: Sage Publications Inc.

Marshall, S. (2009). *Cities, design & evolution*. London/New York: Routledge.

Martignon, L., & Hoffrage, U. (2002). Fast, frugal, and fit: Simple heuristics for paired comparison. *Theory and Decision, 52*(1), 29-71.

Martignon, L., Vitouch, O., Takezawa, M., & Forster, M. R. (2003). Naive and yet enlightened: From natural frequencies to fast and frugal decision trees. In D. Hardman, & L. Macchi (Eds.), *Thinking: Psychological perspectives on reasoning, judgment and decision making* (pp. 189-211). Chichester: Wiley Online Library.

Martin, R., & Sunley, P. (2006). Path dependence and regional economic evolution. *Journal of Economic Geography, 6*(4), 395-437.

Martin, R. (2008). Response to paper "Systems thinking" by D. Cabrera et al.: Systems thinking from a critical systems perspective. *Evaluation and Program Planning, 31*(3), 323-325.

Mathews, K. M., White, M. C., & Long, R. G. (2004). Why study the complexity sciences in the social sciences? *Human Relations, 52*(4), 439-462.

McCray, G. E., Purvis, R. L., & McCray, C. G. (2002; 2002). Project management under uncertainty: The impact of heuristics and biases. *Project Management Journal, 33*(1), 49-49-57.

McKenna, R. J., & Martin-Smith, B. (2005). Decision making as a simplification process: New conceptual perspectives. *Management Decision, 43*(5/6), 821-836.

Meadows, D. H. (1982). Whole earth models and systems. *CoEvolution Quarterly,* 98-108.

Meadows, D. H. (2008). In Wright D. (Ed.), *Thinking in systems: A primer.* White River Junction VT: Chelsea Green.

Merali, Y., & Allen, P. (2011). Complexity and systems thinking. In P.

Allen, S. Maguire & B. McKelvey (Eds.), *The Sage handbook of complexity and management* (pp. 31-52). London: Sage.

Midgley, G., Munlo, I., & Brown, M. (1998). The theory and practice of boundary critique: Developing housing services for older people. *The Journal of the Operational Research Society, 49*(5), 467-478.

Miguel Pina, e. C., & Rego, A. (2010). Complexity, simplicity, simplexity. *European Management Journal, 28*(2), 85-94.

Miller, K. C. (2009). The limitations of heuristics for political elites. *Political Psychology, 30*(6), 863-894.

Mileti, D. (1999). *Disasters by design: A reassessment of natural hazards in the United States*. Washington, DC.: Joseph Henry Press.

Mischen, P. A., & Jackson, S. K. (2008). Connecting the dots: applying complexity theory, knowledge management and social network analysis tot policy implementation. *Public Administration Quarterly, 32*(3), 314-338.

Mitchell, S. (2009). Complexity and explanation in the social sciences. In C. Mantzavinos (Ed.), *Philosophy of the social sciences: Philosophical theory and scientific practice* (pp. 130-145). Cambridge: Cambridge: Cambridge University Press.

Mjøset, L. (2009). The contextualist approach to social science methodology. In D. S. Byrne, & C. C. Ragin (Eds.), *The Sage handbook of case-based methods* (pp. 39-68). London: Sage.

Monroe, K. R., & Epperson, C. (1994). "But what else could I do?" Choice, identity and a cognitive-perceptual theory of ethical political behavior. *Political Psychology*, , 201-226.

Morçöl, G. (2002). *A new mind for policy analysis: Toward a post-newtonian and postpositivist epistemology and methodology* Westport, CT: Praeger Publishers.

Morçöl, G. (2012). *A complexity theory for public policy*. Routledge: New York

Morçöl, G., & Dennard, L. F. (1997). Learning from the natural sciences. *Emergence: Complexity and Organization, 7*(1)

Morçöl, G., & Ivanova, N. P. (2010). Methods taught in public policy programs: Are quantitative methods still prevalent? *Journal of Public Affairs Education*, , 255-277.

Morçöl, G., & Wachhaus, A. (2009). Network and complexity theories: A comparison and prospects for a synthesis. *Administrative Theory & Praxis, 31*(1), 44-58.

Moreno-Peñaranda, R., & Kallis, G. (2010). A coevolutionary understanding of agroenvironmental change: A case-study of a rural community in Brazil. *Ecological Economics, 69*(4), 770-778.

Morton, A. (1993). Heuristics and counterfactual self-knowledge. *Behavioral and Brain Sciences, 16*(01), 63.

Murray, P. J. (2003). So what's new about complexity? *Systems Research and Behavioral Science, 20*, 409-417.

Nabi, R. L. (2003). Exploring the framing effects of emotion do discrete emotions differentially influence information accessibility, information seeking, and policy preference? *Communication Research, 30*(2), 224-247.

Nelson, R. R. (2006). Evolutionary social science and universal Darwinism. *Journal of Evolutionary Economics, 16*(5), 491-510.

Nelson, R. R., & Winter, S. G. (1982). *An evolutionary theory of economic change*. Cambridge: Harvard University Press.

Newell, B. R., Paton, H., Hayes, B. K., & Griffiths, O. (2010). Speeded induction under uncertainty: The influence of multiple categories and feature conjunctions. *Psychonomic Bulletin & Review, 17*(6), 869-874.

Norgaard, R. B. (1984). Coevolutionary development potential. *Land Economics, 60*(2), 160-173.

Norgaard, R. B. (1994). *Development betrayed; the end of progress and a coevolutioary revisioning of the future*. London, New York: Routledge.

Norgaard, R. B. (1995). Beyond materialism: A coevolutionary reinterpration. *Review of Social Economy, 53*(4), 475-486.

Oates, W. E., Howrey, E. P., & Baumol, W. J. (1971). The analysis of public policy in dynamic urban models. *Journal of Political Economy, 79*(1), 142-153.

Odum, E. P. (1971). *Fundamentals of ecology*. Philadelphia: W. B. Saunders Company.

O'Toole Jr, L. J. (1997). Treating networks seriously: Practical and research-based agendas in public administration. *Public Administration Review*, , 45-52.

Otter, H. S. (2000). *Complex adaptive land use systems; an interdisciplinary approach with agent-based models*. Delft: Eburon.

Page, S. E. (2008). Uncertainty, difficulty, and complexity. *Journal of Theoretical Politics, 20*(2), 115-149.

Parsons, T. (1951). *The social system*. London: Routledge & Kegan Paul Ltd.

Parsons, W. (1995). *Public policy; an introduction to the theory and practice of policy analysis*. Cheltenahm: Edward Elgar Publishing Limited.

Patricia J., R. (2008). Response to paper "Systems thinking" by D.

Cabrera et al.: Is it systems thinking or just good practice in evaluation? *Evaluation and Program Planning, 31*(3), 325-326.

Peltzman, S. (1975). The effects of automobile safety regulation. *Journal of Political Economy, 83*(4), 677.

Peterson, S. A. (1985). Neurophysiology, cognition, and political thinking. *Political Psychology, 6*(3), pp. 495-518.

Phelan, S. E. (1999). A note on the correspondence between complexity and systems theory. *Systemic Practice and Action Research, 12*(3), 237-245.

Pierre, J. (2005). Comparative urban governance. *Urban Affairs Review, 40*(4), 446-462.

Pierson, P. (2000). Increasing returns, path dependence and the study of politics. *American Political Science Review, 94*(2), 251-267.

Plutynski, A. (2008). The rise and fall of the adaptive landscape? *Biology and Philosophy, 23*(5), 605-623.

Pollitt, M. G., & Smith, A. S. J. (2002). The restructuring and privatisation of British rail: Was it really that bad? *Fiscal Studies, 23*(4), 463-502.

Porter, T. B. (2006). Coevolution as a research framework for organizations and the natural environment. *Organization & Environment, 19*(4), 479-504.

Portugali, J. (1997). Self-organizing cities. *Futures, 29*(4-5), 353-380.

Pressman, J. L., & Wildavsky, A. B. (1984). *Implementation: How great expectations in Washington are dashed in Oakland: Or, why it's amazing that federal programs work at all, this being a saga of the economic development administration as told by two sympathetic observers who seek to build morals on a foundation of ruined hopes* University of California Press.

Prigogine, I. (1997). *The end of certainty: Time, chaos, and the new laws of nature.* New York: Free Press.

Prigogine, I., & Stengers, I. (1984). *Order out of chaos; man's new dialogue with nature.* Toronto: Bantam Books.

Quade, E. S. (1975). *Analysis for public decisions.* New York: American Elsevier.

Ragin, C. C. (1987). *The comparative method: Moving beyond qualitative and quantitative strategies.* Los Angeles/London: University of California Press.

Ragin, C. C. (1999). Using qualitative comparative analysis to study causal complexity. *Health Services Research, 34*(5), 1225-1239.

Ragin, C. C. (2000). *Fuzzy-set social science.* Chicago: University of Chicago Press.

Ragin, C. C. (2005). *From fuzzy sets to crisp truth tables.* Tucson AZ:

Department of Sociology, University of Arizona. Retrieved from http://www.compasss.org/files/WPfiles/Raginfztt_April05.pdf
Ragin, C. C. (2008). *Redesigning social inquiry: Fuzzy sets and beyond*. Chicago/London: University of Chicago Press.
Ragin, C. C., Shulman, D., Weinberg, A., & Gran, B. (2003). Complexity, generality, and qualitative comparative analysis. *Field Methods, 15*(4), 323-340.

Rammel, C., & Bergh, J. C. J. M. v. d. (2003). Evolutionary policies for sustainable development: Adaptive flexibility and risk minimising. *Ecological Economics, 47*(2 - 3), 121-133.
Rammel, C., Hinterberger, F., & Bechtold, U. (2004). *Governing sustainable development; a co-evolutionary perspective on transitions and change.* Vienna: Governance for Sustainable Development.
Read, D., & Grushka-Cockayne, Y. (2011). The similarity heuristic. *Journal of Behavioral Decision Making, 24*(1), 23-46.
Reed, M., & Harvey, D. L. (1992). The new science and the old: Complexity and realism in the social sciences. *Journal for the Theory of Social Behaviour, 22*(4), 353-380.
Rein, M., & Schön, D. (1996). Frame-critical policy analysis and frame-reflective policy practice. *Knowledge and Policy, 9*(1)
Renn, O. (1992). Risk communication: Towards a rational discourse with the public. *Journal of Hazardous Materials, 29*(3), 465-519.
Rescher, N. (1995). *Luck: The brilliant randomness of everyday life.* Pittsburgh, PA: University of Pittsburgh Press.
Rescher, N. (1998). *Complexity. A philosophical overview.* New Brunswick, New Jersey: Transaction Publishers.
Reynolds, C. W. (1987). Flocks, herds and schools: A distributed behavioral model. *ACM SIGGRAPH Computer Graphics, , 21.* (4) pp. 25-34.
Rhodes, M. L. (2008). Complexity and emergence in public management: The case of urban regeneration in Ireland. *Public Management Review, 10*(3), 361-379.
Rhodes, M. L., & Donnelly-Cox, G. (2008). Social entrepreneurship as a performance landscape: The case of 'Front line'. *Emergence Complexity and Organization, 10*(3), 35-50.
Rhodes, R. A. W. (1988). *Beyond Westminster & Whitehall.* London: Taylor & Francis.
Rhodes, R. A. W. (1997). *Understanding governance: Policy networks, governance, reflexivity and accountability.* Buckingham: Open University Press.
Richard, H. (2008). Response to paper "Systems thinking" by D.

Cabrera et al.: A tool for implementing DSRP in programme evaluation. *Evaluation and Program Planning, 31*(3), 331-333.

Richardson, K. A. (2007). Complexity, information and robustness. *International Workshop on Complexity and Organizational Resilience,* Pohnpei. pp. 176.

Richardson, K. A., & Lissack, M. R. (2001). On the status of boundaries, both natural and organizational: A complex systems perspective. *Emergence, Complexity and Organization, 3*(4), 32-49.

Rihoux, B. (2003). Bridging the gap between the qualitative and quantitative worlds? A retrospective and prospective view on qualitative comparative analysis. *Field Methods, 15*(4), 351-365.

Rihoux, B., & Lobe, B. (2009). The case for qualitative comparative analysis (QCA): Adding leverage for thick cross-case comparison. In D. S. Byrne, & C. C. Ragin (Eds.), *The sage handbook of case-based methods* (pp. 222-242). London: Sage.

Rihoux, B., & Ragin, C. C. (Eds.). (2009). *Configurational comparative methods: Qualitative comparative analysis (QCA) and related techniques*. London: Sage.

Rihoux, B., & Ragin, C. C. (2009). Introduction. In B. Rihoux, & C. C. Ragin (Eds.), *Configurational comparative methods: Qualitative comparative analysis (QCA) and related techniques* (pp. xvii-xxv). London: Sage.

Rilling, J. K., & Sanfey, A. G. (2011). The neuroscience of social decision-making. *Annual Review of Psychology, 62*, 23-48.

Room, G. (2011). *Complexity, institutions and public policy: Agile decision-making in a turbulent world* . Cheltenham: Edgar Elgar.

Room, G. (2011). Social mobility and complexity theory: Towards a critique of the sociological mainstream. *Policy Studies, 32*(2), 109-126.

Rose, A. (2007). Economic resilience to natural and man-made disasters: Multidisciplinary origins and contextual dimensions. *Environmental Hazards, 7*(4), 383–398.

Rosenhead, J. (1998). *Complexity theory and management practice* (scientific paper no. LSEOR 98.25). London: Department of Operational Research, London School of Economics.

Ross, L., & Nisbett, R. E. (1991). *The person and the situation*. Philadelphia: Temple University Press.

Ross, S. B. (1977). On the mode of action of central stimulatory agents. *Acta Pharmacologica Et Toxicologica, 41*(4), 392-396.

Ruijgrok, E. C. M. (2000). Valuation of nature and environment : An overview of Dutch valuation studies. Vrije Universiteit Amsterdam).

Sabatier, P., & Jenkins-Smith, H. C. (1993). *Policy change and learning: An advocacy coalition approach*. Boulder: Westview Press.

Sabini, J., Siepmann, M., & Stein, J. (2001). The really fundamental attribution error in social psychological research. *Psychological Inquiry, 12*(1), 1-15.

Safarzynska, K., & Bergh, J. C. J. M. (2010). Evolving power and environmental policy: Explaining institutional change with group selection. *Ecological Economics, 69*(4), 743-752.

Sanderson, I. (2000). Evaluation in complex systems. *Evaluation, 6*(4), 433-454.

Sanderson, S. K. (1990). *Social evolutionism. A critical history*. Cambridge, M.A.: Blackwell.

Savage, L. J. (1954). *The foundations of statistics*. New York: John Wiley and Sons.

Sayer, A. (1992). *Method in social science: A realist approach*. London: Routledge.

Sayer, A. (2000). *Realism and social science*. London: Sage.

Schaap, L. (1997). Bestuurskunde als bestudering van sociale systemen. 'Theorie der Verwaltungswissenshaft' en ander werk van Niklas Luhmann. *Bestuurskunde, 6*(6), 277-290.

Schaap, L. (1997). *Op zoek naar prikkelende overheidssturing: Over autopoiese, zelfsturing en stadsprovincie*. Delft: Eburon.

Scheffer, M., & Nes, E. H. (2007). Shallow lakes theory revisited: Various alternative regimes driven by climate, nutrients, depth and lake size. *Hydrobiologia, 585*, 455-466.

Scheffer, M., & Westley, F. R. (2007). The evolutionary basis of rigidity: Locks in cells, minds, and society. *Ecology and Society, 12*(2), 36.

Scheffer, M., & Carpenter, S. R. (2003). Catastrophic regime shifts in ecosystems: Linking theory to observation. *Trends in Ecology and Evolution, 18*(12), 648-656.

Scheffer, M., Carpenter, S., Foley, J. A., Folke, C., & Walker, B. (2001). Catastrophic shifts in ecosystems. *Nature, 413*, 591-596.

Schelling, T. C. (1978). *Micromotives and macrobehavior*. New York: W.W. Norton & Co.

Schie, N. v., Edelenbos, J., & Gerrits, L. M. (2010). Organizing interfaces between government institutions and interactive governance. *Policy Sciences, 43*(1), 73-94.

Schipper, D., & Gerrits, L. (2012). On the origin of the MRA. A coevolutionary analysis of the life-and-death cycles of urban governance in the Amsterdam Metropolitan Region. *ASPA Annual Work Conference*, Las Vegas.

Schneider, C. Q., & Wagemann, C. (2010). Standards of good practice in qualitative comparative analysis (QCA) and fuzzy sets. *Comparative Sociology, 9*(3), 397-418.

Schneider, S. C. (1987). Managing boundaries in organizations. *Political Psychology, 8*(3), pp. 379-393.

Schön, D. A., & Rein, M. (1994). *Frame reflection: Toward the resolution of intractable policy controversies*. New York: Basic Books.

Schwenk, C. R. (1984). Cognitive simplification processes in strategic decision-making. *Strategic Management Journal, 5*(2), 111-128.

Seidl, D., & Becker, K. H. (Eds.). (2005). *Niklas Luhmann and organization studies* . Copenhagen: Copenhagen Business School Press.

Seidl, D. (2005). The basic concepts of Luhmann's theory of social systems. In D. Seidl, & K. H. Becker (Eds.), *Niklas Luhmann and organization studies* (pp. 21-53). Malmö: Liber.

Sementelli, A. (2007). Distortions of progress. *Administration & Society, 39*(6), 740-760.

Senge, P. M. (1990). *The fifth discipline: The art and practice of the learning organisation*. New York: Currency/Doubleday.

Shafir, E., Simonson, I., & Tversky, A. (1993). Reason-based choice. *Cognition, 49*(1-2), 11-36.

Shah, A. K., & Oppenheimer, D. M. (2008). Heuristics made easy: An effort-reduction framework. *Psychological Bulletin, 134*(2), 207.

Shanteau, J., & Thomas, R. P. (2000). Fast and frugal heuristics: What about unfriendly environments? *Behavioral and Brain Sciences, 23*(05), 762.

Sharkansky, I. (2002). *Politics and policymaking: In search of simplicity*. Boulder: Lynne Rienner Publishers.

Sibeon, R. (1999). Anti-reductionist sociology. *Sociology, 33*(2), 317-334.

Siggelkow, N., & Levinthal, D. A. (2003). Temporarily divide to conquer: Centralized, decentralized and, reintegrated organizational approaches to exploration and adaptation. *Organization Science, 14*(6), 650-669.

Silva, S. T., & Teixeira, A. A. C. (2009). On the divergence of evolutionary research paths in the past 50 years: A comprehensive bibliometric account. *Journal of Evolutionary Economics, 19*(5), 605-642.

Simon, H. A. (1956). Rational choice and the structure of the environment. *Psychological Review; Psychological Review, 63*(2), 129.

Simon, H. A. (1962). The architecture of complexity. *Proceedings of the American Philosophical Society, 106*(6), 467-482.

Simon, H. A. (1991). Bounded rationality and organizational learning.

Organization Science, 2(1), 125-134.

Skyrms, B. (1996). *Evolution of the social contract.* Cambridge: University Press.

Slovic, P., Fischhoff, B., & Lichtenstein, S. (1979). Rating the risks. *Environment: Science and Policy for Sustainable Development, 21*(3), 14-39.

Slovic, P., Finucane, M. L., Peters, E., & MacGregor, D. G. (2007). The affect heuristic. *European Journal of Operational Research, 177*(3), 1333-1352.

Smets, P., & Salman, T. (2008). Countering urban segregation: Theoretical and policy innovations from around the globe. *Urban Studies, 45*(7), 1307.

Smith, L. L. (2002). Economies and markets as complex systems. *Business Economics, 37*(1), 46-53.

Smith, T. S., & Stevens, G. T. (1996). Emergence, self-organisation, and social interaction: Arousal dependent structure in social systems. *Sociological Theory, 14*(2), 131-153.

Smithson, M., & Verkuilen, J. (2006). *Fuzzy set theory: Applications in the social sciences.* Thousand Oaks: Sage.

Sokal, A., & Bricmont, J. (1999). *Fashionable nonsense; postmodern intellectuals' abuse of science.* St. Martins: Picador Press.

Solow, D., & Szmerekovsky, J. G. (2006). The role of leadership: What management science can give back to the study of complex systems. *Emergence Complexity Organization, 8*(4), 52-60.

Sornette, D. (2003). *Why stockmarkets crash: Critical events in complex financial systems.* Princeton, New Jersey: Princeton University Press.

Sotarauta, M., & Srinivas, S. (2006). Co-evolutionary policy processes: Understanding innovative economies and future resilience. *Futures, 38*(3), 312-336.

Stacey, R. D. (2003). *Strategic management and organisational dynamics* (4th ed.). London etc.: FT Prentice Hall.

Stacey, R. D. (Ed.). (2005). *Experiencing emergence in organizations: Local interaction and the emergence of global pattern.* New York: Routledge.

Stacey, R. D., & Griffin, D. (Eds.). (2006). *Complexity and the experience of managing in public sector organizations.* London New York: Routledge.

Sterman, J. D. (2000). *Business dynamics: Systems thinking and modeling for a complex world.* McGraw-Hill: Irwin.

Stiglitz, J. E. (2010). Risk and global economic architecture: Why full financial integration may be undesirable. *NBER Working Paper,*

Stone, D. A. (1997). *Policy paradox: The art of political decision making*. New York: WW Norton.

Strand, R. (2002). Complexity, ideology, and governance. *Emergence, 4*(1), 164-183.

Strogatz, S. H. (1994). *Nonlinear dynamics and chaos. with applications to physics, biology, chemistry and engineering*. Reading, MA: Perseus Books Publishing.

Suedfeld, P., & Granatstein, J. L. (1995). Leader complexity in personal and professional crises: Concurrent and retrospective information processing. *Political Psychology, 16*(3), pp. 509-522.

Sunstein, C. R. (2005). Moral heuristics. *Behavioral and Brain Sciences, 28*(04), 531.

Swyngedouw, E., Moulaert, F., & Rodriguez, A. (2002). Neoliberal urbanization in Europe: Large–Scale urban development projects and the new urban policy. *Antipode, 34*(3), 542-577.

Teisman, G. R. (2005). *Publiek management op de grens van chaos en orde: Over leidinggeven en organiseren in complexiteit* (first ed.). Den Haag: Sdu Uitgevers.

Teisman, G. R., Westerveld, E., & Hertogh, M. J. C. M. (2009). Appearances and sources of process dynamics: The case of infrastructure development in the UK and the netherlands. In G. R. Teisman, M. W. Van Buuren & L. M. Gerrits (Eds.), *Managing complex governance systems: Dynamics, self-organization and coevolution in public investments* (pp. 56-75). New York: Routledge.

Tsoukas, H., & Hatch, M. J. (2001). Complex thinking, complex practice: The case for a narrative approach to organizational complexity. *Human Relations, 54*(8), 979-1013.

Turner, B. S. (2005). Talcott Parsons's sociology of religion and the expressive revolution. *Journal of Classical Sociology, 5*(3), 303-318.

Tushman, M., & Anderson, P. (1986). Technological discontinuities and organizational environments. *Administrative Science Quarterly, 31*(3), 439-465.

Tversky, A., & Kahneman, D. (1973). Availability: A heuristic for judging frequency and probability. *Cognitive Psychology, 5*(2), 207-232.

Tversky, A., & Kahneman, D. (1983). Extensional versus intuitive reasoning: The conjunction fallacy in probability judgment. *Psychological Review, 90*(4), 293.

Ulrich, W. (2005). A brief introduction to critical systems heuristics (CSH). *Ecosensus*

Uprichard, E., & Byrne, D. (2006). Representing complex places: A narrative approach. *Environment and Planning A, 38*(4), 665.

Van Ast, J., van Schie, N., Edelenbos, J., & Gerrits, L. (2010). Arnemuiden aan de Arne? *H2O, 41*(21), 17.

Van Gils, M., Gerrits, L. M., & Teisman, G. R. (2009). Non-linear dynamics in port systems: Change events at work. In G. R. Teisman, M. W. Van Buuren & L. M. Gerrits (Eds.), *Managing complex governance systems: Dynamics, self-organization and coevolution in public investments* (pp. 76-96). New York: Routledge.

Van Hiel, A., & Mervielde, I. (2003). The measurement of cognitive complexity and its relationship with political extremism. *Political Psychology, 24*(4), 781-801.

Van Schie, N. (2010). *Co-valuation of water: An institutional perspective on valuation in spatial water management.* Rotterdam: Erasmus University Rotterdam.

Varela, F. J., Maturana, H. R., & Uribe, R. (1974). Autopoiesis: The organization of living systems, its characterization and a model. *Biosystems, 5*, 187-196.

Vella, A., & Morad, M. (2011). Taming the metropolis: Revisiting the prospect of achieving compact sustainable cities. *Local Economy, 26*(1), 52-59.

Venkatraman, N. (1989). The concept of fit in strategy research: Toward verbal and statistical correspondence. *Academy of Management Review,* , 423-444.

Vertzberger, Y. Y. I. (1997). Collective risk taking: The decision making group. In P. Hart 't, E. K. Stern & B. Sundelius (Eds.), *Beyond groupthink, political group dynamics and foreign policy-making* (pp. 275-336). Ann Arbor: The University of Michigan press.

Vertzberger, Y. Y. I. (1995). Rethinking and reconceptualizing risk in foreign policy decision-making: A sociocognitive approach. *Political Psychology, 16*(2), pp. 347-380.

Verweij, S. & Gerrits, L. M. (2012). Evaluating transportation infrastructure projects: understanding and researching complexity with qualitative comparative analysis. *Evaluation (forthc.)*

Vesterby, V. (2008). Measuring complexity: Things that go wrong and how to get it right. *Emergence: Complexity and Organization, 10*(2), 90-102.

Vis, B. (2007). States of welfare or states of workfare? welfare state restructuring in 16 capitalist democracies, 1985-2002. *Policy & Politics, 35*(1), 105-122.

Von Bertalanffy, L. (1968). *General system theory. foundations, development, applications*. New York: George Braziller.

Vromans, M. J. C. M. (2005). Reliability of railway systems. (PhD, Erasmus Research Institute of Management).

Wagenaar, H. (1997). Beleid als fictie: Over de rol van verhalen in de bestuurlijke praktijk. *Beleid & Maatschappij, 24*(1), 7-21.

Wagenaar, H. (1997). Verhalen in de beleidspraktijk. *Beleid & Maatschappij, 24*(1), 2-6.

Wagenaar, H. (2007a). Governance, complexity and democratic participation: How citizens and public officials harness the complexities of neighborhood decline. *The American Review of Public Administration, 37*(1), 17-50.

Wagenaar, H. (2007b). Philosophical hermeneutics and policy analysis: Theory and effectuations. *Critical Policy Analysis, 1*(4), 311-341.

Wagenaar, H., & Cook, S. D. N. (2003). Understanding policy practices: Action, dialectic and deliberation in policy analysis. In M. A. Hajer, & H. Wagenaar (Eds.), *Deliberative policy analysis: Understanding governance in the network society* (pp. 139-171). Cambridge: Cambridge University press.

Waldman, J. (2007). Thinking systems need systems thinking. *Systems Research and Behavioral Science, 24*(3), 271-284.

Waldrop, M. M. (1992). *Complexity; the emerging science at the edge of order and chaos*. New York: Touchstone.

Walker, B., & Meyers, J. A. (2004). Thresholds in ecological and Social–Ecological systems: A developing database. *Ecology and Society, 9*(2), 1-16.

Waring, T. M. (2010). New evolutionary foundations: Theoretical requirements for a science of sustainability. *Ecological Economics, 69*(4), 718-730.

Weiss, R. S. (1995). *Learning from strangers: The art and method of qualitative interview studies*. New York: Free Press.

Williams, M. (2009). Social objects, causality and contingent realism. *Journal for the Theory of Social Behaviour, 39*(1), 1-18.

Williams, M. (2011). Contingent realism - abandoning necessity. *Social Epistemology, 25*(1), 37-56.

Williamson, O. E. (1998). The institutions of governance. *The American Economic Review, 88*(2, Papers and Proceedings of the Hundred and Tenth Annual Meeting of the American Economic Association), pp. 75-79.

Wimsatt, W. C. (2000). Heuristics refound. *Behavioral and Brain Sciences, 23*(05), 766.

Winkielman, P., Zajonc, R. B., & Schwarz, N. (1997). Subliminal affective priming resists attributional interventions. *Cognition & Emotion, 11*(4), 433-465.

Witt, U. (2003). Economic policy making in evolutionary perspective. *Journal of Evolutionary Economics, 13*(2), 77-94.

Wright, S. (1932). The roles of mutation, inbreeding, crossbreeding and selection in evolution. *Proceedings of the Sixth International Congress on Genetics, , 1.* (6) pp. 356-366.

Wuisman, J. J. J. M. (2005). The logic of scientific discovery in critical realist social scientific research. *Journal of Critical Realism, 4*(2), 366-394.

Yamagishi, T., Terai, S., Kiyonari, T., Mifune, N., & Kanazawa, S. (2007). The social exchange heuristic: Managing errors in social exchange. *Rationality and Society, 19*(3), 259-291.

Zajonc, R. B. (1980). Feeling and thinking: Preferences need no inferences. *American Psychologist, 35*(2), 151.

Zia, A., & Koliba, C. Climate change governance and accountability: Dilemmas of performance measurement in complex governance networks.

Zuidema, C., & De Roo, G. (2004). Complexiteit als planologisch begrip. *Rooilijn, ,* 485-490.

INDEX PUNCHING CLOUDS

Lightning Source UK Ltd.
Milton Keynes UK
UKOW03f0744291213

223687UK00017B/1182/P

9 781938 158001